Textile Waste Recycling and Sustainable Textile Product Development

This book covers various aspects involved in the recycling of textile waste into value-added products and its impact on the environment and circular economy. It discusses current challenges of recycling technologies and the scope for research and innovation in new technologies, offering a comprehensive and inclusive approach aimed at advancing sustainable practices within the textile industry.

Key features:

- Exclusively covers recycling and sustainability from different aspects, including new technologies and products
- Addresses a pressing question for environmental management and resource recovery
- Provides a holistic view of textile waste management, covering everything from waste collection and sorting to innovative product development
- Showcases examples of industrial textiles developed using recycled textiles, including detailed case studies and design principles
- Delves into various recycling methods for natural, synthetic, blends, and composites, thereby covering all aspects of textile materials

This book is aimed at researchers and graduate students in textile engineering, waste recycling, and sustainability analysis.

Textile Waste Recycling and Sustainable Textile Product Development

G. Thilagavathi
Neelakrishnan Subramanyan
Rangam Rajkhowa
M. Jaya Bharathi

CRC Press
Taylor & Francis Group
Boca Raton London New York

CRC Press is an imprint of the
Taylor & Francis Group, an **informa** business

Designed cover image: G. Thilagavathi, Neelakrishnan Subramanyan, Rangam Rajkhowa and M. Jaya Bharathi.

First edition published 2026
by CRC Press
2385 NW Executive Center Drive, Suite 320, Boca Raton FL 33431

and by CRC Press
4 Park Square, Milton Park, Abingdon, Oxon, OX14 4RN

CRC Press is an imprint of Taylor & Francis Group, LLC

ISBN: 978-1-032-79687-1 (hbk)
ISBN: 978-1-032-82278-5 (pbk)
ISBN: 978-1-003-50383-5 (ebk)

DOI: 10.1201/9781003503835

Typeset in Times
by SPi Technologies India Pvt Ltd (Straive)

Contents

Foreword

The global textile industry is witnessing increasing environmental and social challenges related to textile production, consumption, and disposal. This book addresses textile waste recycling and sustainable product development geared towards circularity.

The authors have provided a structured and comprehensive understanding of textile waste generation, classification, and the recycling technologies currently employed to recover value from end-of-life textiles. They have systematically explored different recycling methods along with the associated challenges. This book provides readers with both a conceptual framework and practical insights through several case studies taken from experimental and product development activities from centres of excellence and industry practices. The authors have also integrated a chapter on digitalization tools such as fibre identification and digital product passports, demonstrating the importance of technology in achieving sustainability.

The emphasis it lays on principles and applications such as thermal and acoustic insulation, oil sorbents, construction, packaging, agro-textiles, and filtration should be of great value to textile technologists, industries, researchers, practitioners, and policymakers. It will also serve as a reference textbook for students seeking a thorough understanding of sustainable textile technology and management.

I hope this book will inspire further research, innovation, and responsible practice in the textile industry, helping to pave the way towards a truly circular textile economy.

Dr. Prasad Potluri
Professor of Robotics and Textile Composites
Director of Northwest Composites Centre
University of Manchester, UK

Preface

The global textile industry stands at a crossroads where innovation must align with sustainability. The increasing accumulation of textile waste, coupled with the depletion of natural resources, has created an urgent need for circular approaches that minimize waste, extend product lifecycles, and promote resource efficiency. This book, *Textile Waste Recycling and Sustainable Textile Product Development*, aims to provide a comprehensive understanding of the principles, technologies, and applications driving this transformation towards a circular textile economy.

This book begins with an exploration of the circular economy, contrasting it with traditional linear models and examining the environmental challenges associated with the textile sector. It highlights the pressing issues arising from production, consumption, and disposal patterns while identifying pathways to achieve circularity. The discussion then extends to the generation and categorization of textile waste, presenting the diverse waste streams originating from various textile processes and their potential as valuable secondary raw materials.

A detailed analysis of recycling technologies forms the core of this book. It encompasses mechanical, thermal, and chemical recycling processes, alongside advanced depolymerization techniques and strategies for managing fibre-reinforced composites. The challenges posed by dyes and finishing agents are also addressed, emphasizing the need for integrated waste management solutions. With the rise of Industry 5.0, a dedicated chapter discusses digital tracking, fibre identification, and separation technologies, including the emerging concept of the digital product passport, to enhance recycling efficiency and traceability.

To bridge theory and practice, the book discusses case studies demonstrating industrial sustainable product development and suitable applications. These examples illustrate innovative approaches in fibre sorting, product development methods, and closed-loop systems that contribute to a sustainable textile ecosystem across diverse sectors ranging from thermal and acoustic insulation to packaging, construction, and agro-textiles, thereby redefining waste as a resource.

This book serves as a valuable reference for students of textile technology, fashion design and technology, and other disciplines of engineering, researchers, industry professionals, and policymakers dedicated to building a sustainable textile future through responsible innovation and resource-conscious design.

Acknowledgements

We express our sincere gratitude to the Management of **PSG College of Technology**, Coimbatore, India and the Principal for their continuous encouragement, support, and for providing the excellent facilities that enabled us to successfully complete this book, *Textile Waste Recycling and Sustainable Textile Product Development.*

We also wish to thank our Centre of Excellence for Industrial Textiles—**PSGTECHS COE INDUTECH**, Neelambur, India for their invaluable support in facilitating our research work through access to advanced testing and analytical facilities.

We extend our sincere thanks to **Deakin University**, Australia, for their active collaboration in this work through joint research efforts, knowledge exchange, and valuable technical contributions that enriched the overall quality of this publication.

We gratefully acknowledge the financial and technical support provided by the **SPARC** (Scheme for Promotion of Academic and Research Collaboration) project under the Ministry of Human Resource Development (MHRD), Government of India. The project greatly facilitated our collaborative research activities, exchange of expertise, and access to advanced resources, which strengthened the academic and practical aspects of this book.

We extend our heartfelt thanks to **Shree Renga Polyester Pvt Ltd**, Karur, India for their valuable industrial collaboration and for providing the case studies featured in this book. Their contributions offered us real-world insights into textile recycling practices and sustainable product development, bridging the gap between academic study and industrial implementation.

Finally, we thank our colleagues, research scholars, and students from the Department of Textile Technology, PSG College of Technology, for their constant support, constructive discussions, and encouragement during the preparation of this book.

Dr. G. Thilagavathi
Dr. Neelakrishnan Subramanyan
Dr. Rangam Rajkhowa
M. Jaya Bharathi

About the Authors

G. Thilagavathi is Professor and Head of Textile Technology at PSG College of Technology since 2010, with over 30 years of experience in teaching, research, training, and administration. She also headed Fashion Technology (2008–2013). An eminent researcher in Textile Quality Evaluation, Functional and Industrial Textiles, she currently manages sponsored projects exceeding Rs. 50 crores.

Dr Thilagavathi has published over 155 research papers with an h-index of 34 and an i-10 index of 85. She has organized six international conferences (Indutech) and many workshops and National Conference. She has delivered more than 50 research papers in international conferences and visited countries like Germany, UK, Singapore, China.

She is a recipient of several prestigious awards, including the Dr. R. Sundaramoorthy Award for Outstanding Academician of PSG College of Technology (2012–2016) and the Eminent Engineer Award from the Institution of Engineers (India), Kolkata. She is a Fellow of the Institution of Engineers and serves as Chairman of the Textile Division (IEI). Dr. Thilagavathi holds key positions in academic and policy bodies, including BIS—Technical Committee on Industrial Textiles (TXD 33), IIHT, Central Silk Board, AICTE, Ministry of Textiles (MoT), and various governing and academic councils. Under her leadership, the department received the AICTE-CII Award (2015) and long-term NBA accreditation.

She established a Centre of Excellence in Industrial Textiles with Ministry of Textiles funding (Rs.25 Crores) and developed several advanced laboratories- Focus Incubation Centre (Rs.2.85 Crores), Antimicrobial Finishing and Evaluation Laboratory, Microencapsulation Laboratory, Fibre Manufacturing Laboratory, Electroactive Textile Laboratory, and NTTM funding for R&D and infrastructure (Rs. 10 crores).

Neelakrishnan Subramanyan has a professional career spanning more than three decades. His previous assignments at Nava Consultants Private Limited and Alliance Automation Systems, USA involving assembly automation systems.

During his service at PSG Tech for the past 17 years, he has been instrumental in the development of vocational programs for the automotive domain with several leading automotive industries, and with setting up of 15 specialized state of the art laboratories in the automotive and textile domains. He was instrumental in setting up of the Partners for the Advancement of Collaborative Engineering (PACE) Laboratory at PSG Tech with support from General Motors India. This involved funding from the PACE partners to the tune of Rs. 1500 Crores. He was also the Co-Principal Coordinator for the technical textiles (industrial and automotive) centre of excellence set up at PSG at a cost of Rs. 45 Crore with funding to the tune of Rs. 25 Crore from the Ministry of Textiles, Government of India.

Dr. Neelakrishnan Subramanyan has published more than 51 research papers and has a h-index of 12 and i-10 index of 19. He has guided ten PhD scholars, with another five of them pursuing their doctoral research and has a patent to his credit. Dr. Neelakrishnan Subramanyan has carried out several consultancy projects in the areas of automotive technologies. Currently the areas of research are design for manufacturing, automotive acoustics, collaborative product design, and Electric and Hybrid vehicles.

He has successfully carried out funded and sponsored research projects funded by the Government and industries. Dr. Neelakrishnan Subramanyan currently heads the Department of Automobile Engineering at PSG College of Technology which focuses on research in the areas of automotive design, emission, electric and hybrid technologies, automotive electronics and automotive functional textiles.

Rangam Rajkhowa is a Professor at the Deakin Institute for Frontier Materials (IFM), Deakin University, Australia. He currently leads the Future Fibres Group, with around 100 researchers within the IFM. He has an extensive research and publication record, with over 100 journal papers, ten full conference papers, and three books. He has also delivered more than 60 conference presentations. He has presented key note and invited presentations in leading international seminars including Gordon Research Conference (GRC) on "Silk Proteins and the Transition to Biotechnologies", International Conference on Industrial Textiles (InduTech 2025, 2020, 2018) and a number of leading universities in Europe, USA and India.

Dr. Rajkhowa holds two Australian provisional patents and five international PCT applications, of which two have been successfully granted in Australia, USA, UK, France, India, China, Japan, South Korea, Canada, and Germany. He has supervised 33 PhD scholars, including 16 as Principal Supervisor, with 15 successful completions to date. He has secured approximately $19 million in research funding, including $ 2.9 million as Lead Chief Investigator, from agencies such as the Australian Research Council, Cotton Incorporated (USA), and industries from Australia, New Zealand and Israel and other sources.

Dr. Rangam received the Anthem Gold Award 2024, selected from 2,000 submissions representing 44 countries. In 2017, he was a recipient of the Global Change Award from the H&M Foundation, receiving €150,000 in funding; his project was selected as one of only five winning concepts from 2,900 applications across 130 countries worldwide. He was also awarded the VC Award for Outstanding Contribution to Industry and Employer Partnerships in 2019. In addition, he has been honored by the Australian Academy of Science with both the Australia–India Early Career Fellowship (2012) and Mid-Career Fellowship (2018), valued at $20.4K and $24K, respectively.

M. Jaya Bharathi serves as an Assistant Professor in the Department of Textile Technology at PSG College of Technology, Coimbatore. She has four years of teaching and research experience in the field of Textile Technology. She earned her Bachelor's degree in Textile Technology from Anna University, Chennai, in 2018, and her Master's degree from PSG College of Technology in 2021. She is currently pursuing her Ph.D. under the guidance of Dr. G. Thilagavathi, focusing on the development of sustainable textile products utilizing natural fibres and biopolymer. Her research interests encompass Textile Quality Evaluation, Protective Textiles, Fibre reinforced composites and Acoustic Textiles. She has published one research paper and two conference proceedings. In addition to her academic and research pursuits, she has actively contributed to institutional activities by coordinating a Faculty Development Programme and an International Conference under the SPARC project.

1 Introduction to Circular Economy and Its Relevance to Textiles

1.1 INTRODUCTION

In today's modern era, textiles have become an integral part of a daily lifestyle of human beings. This growing demand for clothing has given rise to a situation called "fast fashion." The term "fast fashion" was first coined by *The New York Times*, when a fashion brand called "ZARA" entered the market in New York City (1). The strategies of fast fashion are producing more clothing, rapid change in recent trends, marketing at affordable prices for frequent purchases, heavy discounts, online platform sales, and retailing in a short span (see Figure 1.1). While this model has made fashion more accessible to people, it has also had profound environmental and social consequences.

FIGURE 1.1 Strategies of fast fashion.

DOI: 10.1201/9781003503835-1

Rapid production and consumption result in high carbon emissions, excessive water use, and growing textile waste. Globally, the textile industry is considered to be one of the major causes for generation of textile waste, more than 92 million tons each year which is expected to be 134 million tons annually in 2030 (2). This waste represents roughly 12–15% of total municipal solid waste in developed countries and accounts for an estimated 3% of global landfill volume. Most discarded garments are sent to landfills where they decompose slowly and release harmful substances, while a smaller fraction enters informal recycling streams. An equally alarming impact on the environment by fast fashion is the release of microfibres during laundering. Studies revealed that a single garment can release up to 1,900 fibre fragments per wash, contributing to the growing problem of microplastic pollution in oceans and water bodies affecting the aquatic ecosystem (3).

Fast fashion focuses on the linear economy which adheres to the pattern of the "take-make-use-dispose" paradigm. The natural resources were converted into commodities and then disposed of after limited use, resulting in an unsustainable approach and added pressure on conserving the environment. As a solution to these challenges, circular economy (CE) has been projected as a good alternative. In contrast to the linear model, the CE is grounded on the principle of designing out waste and pollution, keeping materials and products in use, and regenerating natural systems. The shift from a linear to a CE model is not only vital for minimizing the environmental footprint of the textile industry but also affords opportunities for innovation, economic development, and attaining Sustainable Development Goals (SDGs).

1.2 LINEAR VS. CIRCULAR ECONOMY

The fundamental difference between the linear and circular models lies in their approach to resource management as shown in Figure 1.2. In a linear economy, a product is made, used, and then disposed of; it is a one-way flow of resources that results in significant waste and environmental degradation. In contrast, CE aims for a closed-loop system designed to minimize waste and maximize resource efficiency by using a product as long as possible. CE is fundamental to a sustainable economic growth model with a net positive impact. This is achieved by durable design, encouraging reuse, resale, and repair to extract maximum value and then recycling at the end when no more use is possible. CE is involved both in the manufacturing phase as well as in the use phase of a product.

According to the definition of the Ellen MacArthur Foundation, CE design reduces or, if possible, eliminates waste during use and disposal phases. It also involves designing pollution-free eco-friendly production of goods through low energy, low resource, and less harmful production processes. CE design principles also support regenerative, biodegradable, or compostable material systems in place of fossil fuel or non-renewable materials which is referred to as Circular Bio-Economy.

Overall, CE benefits the environment, society, and also business through sustainable uses of resources. Environmental benefits come from waste reduction and associated pollution and carbon footprint reduction from a product lifecycle. Business opportunities come from resale, renting platforms, and recycling enterprises. Social benefits come from household savings and involvement in social enterprises dealing with collection of waste, and as well as from a clean environment.

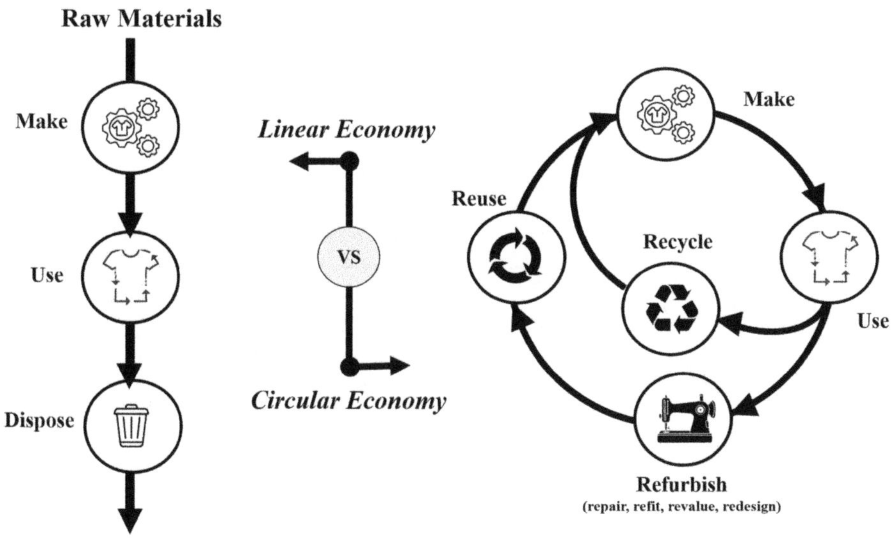

FIGURE 1.2 Linear economy vs. circular economy.

1.3 IMPLEMENTATION OF CIRCULAR ECONOMY PRINCIPLES

In textile and fashion CE, principles of circularity are to be implemented at various stages. Figure 1.3 shows four key areas on which circularity and sustainability rely. Strategies are required for product design, materials selection, textile production, and use and after-use stages of textile products. Success in circularity and sustainability in the textile sector depends on innovations in technology, industry engagements, new business models in all phases of the value chain, regulations, finance, and changing consumer behaviour through concerted efforts.

The transition to a CE is essential to mitigate these impacts, and it is built upon four core principles:

1.3.1 Design to Help Sustainable Production and Circularity

The role of design is pivotal in enabling circularity. A large part of a product's environmental impact is determined at the design stage, making early decisions about fibre choice, construction, selection of accessories, and end-of-life options critical. Design for disassembly ensures products can be taken apart easily for repair or recycling. Avoiding complex fibre blends—such as polyester-cotton mixtures, which are notoriously hard to recycle—can make products more compatible with circular systems. Furthermore, modular or adaptable designs increase longevity by allowing garments to be resized, restyled, or repurposed. Digital tools such as 3D virtual prototyping also reduce sampling waste, cutting textile waste in product development. By embedding circularity into design, companies can transform their products from disposable goods into long-term material assets.

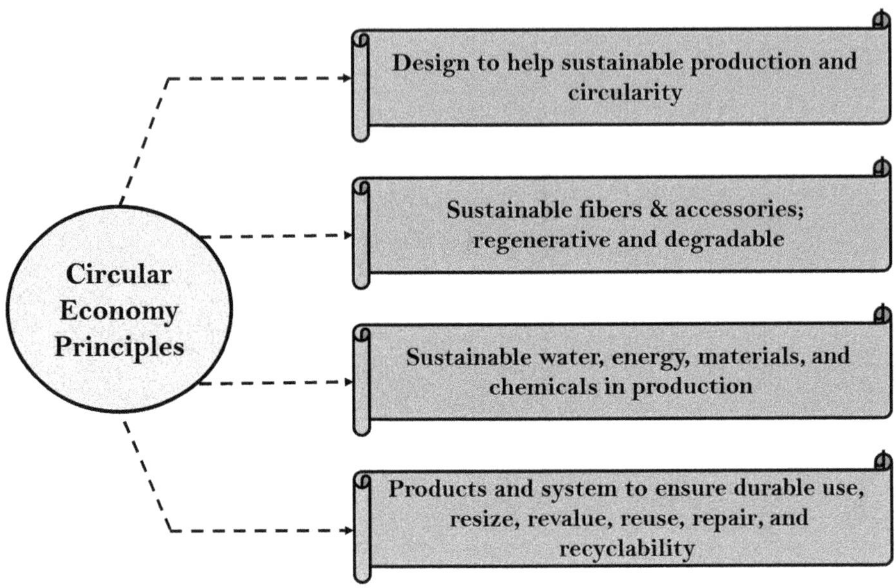

FIGURE 1.3 Principles of circular economy.

1.3.2 Sustainable Fibres and Materials, Regenerative and Degradable Materials

The shift towards regenerative and biodegradable materials is essential to reducing the textile sector's heavy dependence on finite, petroleum-based resources. Currently, over 60% of textiles are produced from synthetic fibres such as polyester, nylon, and acrylic, which are derived from fossil fuels and persist in the environment for centuries (4). In contrast, regenerative materials like cotton and wool improve soil health and support biodiversity. Degradable fibres also mitigate microplastic pollution, where synthetic textiles contribute nearly 35% of global microplastic release into oceans, while natural fibres biodegrade safely (5).

1.3.3 Sustainable Water, Energy, Materials, and Chemicals in Production

The production stage of textiles carries a substantial environmental burden. Textile dyeing and finishing alone account for 20% of global industrial water pollution. Reducing virgin resource use, adopting renewable energy, and eliminating hazardous chemicals can mitigate these impacts. The transition to non-toxic alternatives is critical, as over 8,000 synthetic chemicals are used in textile production, many of which pose risks to both ecosystems and human health (6).

1.3.4 Products and Systems to Ensure Durable Use, Reuse, Repair, and Recyclability

Extending the life of garments is one of the most impactful ways to reduce environmental pressure. Designing for durability means selecting high-quality fibres, strong seams, and modular designs that allow easy repair. Business models such as clothing rental, resale, and upcycling are rapidly gaining traction. Additionally, designing products with mono-materials or easily separable components enhances recyclability. Creating systems that encourage repair and recycling is therefore a cornerstone of circularity. Brands and retailers have begun adopting modular designs, detachable components, and standardized parts to facilitate repair and promote longevity. In addition, resale, rental, and refurbishment platforms are emerging as key enablers of circularity.

Circularity is made possible in large part by consumer behaviour. Waste can be considerably decreased by educating customers about clothing maintenance, repair, resale, and proper disposal. A comprehensive approach to circular textiles combines consumer education, zero-waste principles, repairability, design, and resale or rental models. According to the 2023 report by ThredUp and GlobalData, the global second-hand market is projected to double in size by 2027. It currently represents less than 10% of overall apparel sales, highlighting the infancy of circular systems in textiles (7). By combining design innovation with systemic circular strategies, the textile industry can transition from a linear "take-make-dispose" model to a sustainable, regenerative, and resource-efficient model that supports both economic and environmental goals.

1.4 STATE OF THE CURRENT CLOTHING AND TEXTILE INDUSTRY

CE is highly relevant to textiles today more than ever before due to the rapid growth of consumption from "fast fashion" as briefly discussed earlier. Consumers are lured by rapid changes in design trends and the low price of products, encouraging large purchase responses from consumers. This is facilitated also by increased and rapid production and availability of cheap synthetic fibres, primarily polyester. According to the 2023 Textile Exchange report, polyester accounts for nearly 57% of global fibre production, and its output has more than doubled since 2000 due to low cost and versatility (8). High textile consumption is supported by higher affordability of people, mostly in highly populous and rapidly growing economies like India and China, and the surge of online platforms offering heavy discounts. The global fashion industry now produces around 100 billion garments annually, with China and India being the fastest-growing consumer markets.

Fast fashion leads to a large fraction of unsold and/or unused products being wasted, representing a tremendous loss of economic and material value. It is estimated that one truckload of textiles is landfilled or incinerated every second globally, equating to over 92 million tonnes of textile waste annually (2). More than 80% of textiles leak out of the system at the end of life: they are incinerated, landfilled, or end up in the environment. Textile collection systems are underdeveloped and do not capture all textiles placed on the market. Currently, less than 15% of post-consumer textiles are collected for reuse or recycling, and only about 1% is recycled back into new garments (9). In the current environment, the economics of collection and sorting of discarded textiles happen at a cost, and sorters around the world experience challenges due to manual operations.

Automatic sorting systems are at the early stages of development and remain costly and less efficient. For instance, fibre identification technologies, such as near-infrared (NIR) scanning, can be very expensive to implement, limiting their adoption. Moreover, designing of products is not done keeping in mind the CE, impacting fibre-to-fibre recycling. Besides, reuse platforms are also at their early stages in most countries.

Currently, only about 1% of used textiles are recycled into new garments. When recycling is not feasible, these textiles are often downcycled (see Figure 1.4). About 12% of textile waste is downcycled into products like insulation, cleaning rags, and industrial materials. Downcycling helps reduce raw material demand but often degrades fibre quality, limiting reuse. For example, cotton-polyester blends are challenging to recycle due to complex separation processes, and only 5–10% of blended textiles are currently processed into secondary products.

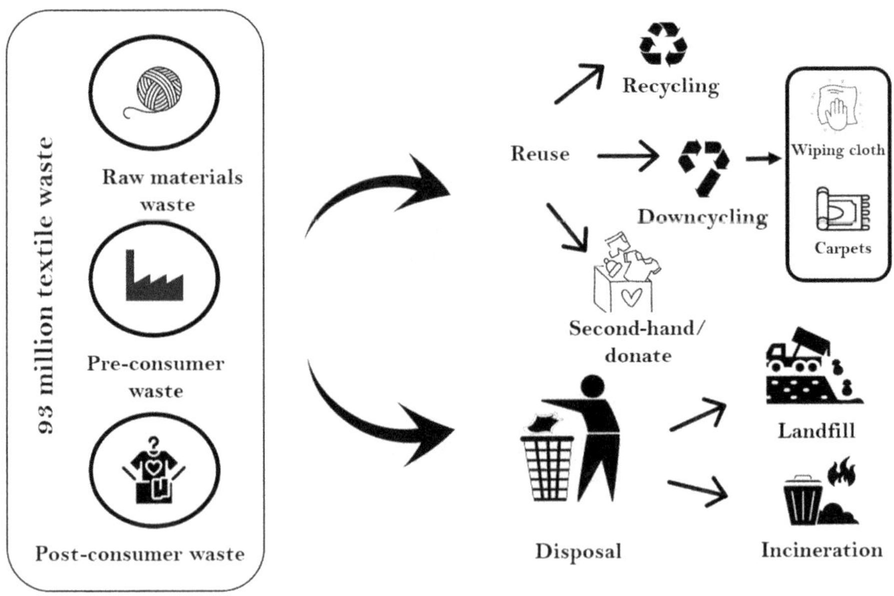

FIGURE 1.4 Journey of textile waste between sustainability and landfill.

1.5 ENVIRONMENTAL IMPACT FROM PROCESSING AND END-OF-LIFE TEXTILES AND MICROFIBRES

A critical and alarming consequence of the current linear model and fast fashion is the widespread microplastic and chemical contamination. The sheer scale of production places an immense strain on natural resources, particularly water. The process of creating a single t-shirt alone consumes approximately 2,700 litres of water, an amount that could sustain one person for 2.5 years of drinking water, underscoring the severe water consumption issue (10).

The washing of synthetic textiles, such as those made from acrylic, polyester, or polypropylene, releases microfibres into the environment. These tiny plastic fragments are a major source of pollution, with an estimated 35% of microplastics found in the ocean originating from the washing of textiles. According to studies, appropriate maintenance can increase clothing life by 25–30% and lower microfibre emissions by up to 80%. This includes washing at lower temperatures, letting clothes air dry, and limiting the frequency of washings. Initiatives spearheaded by brands such as repair workshops, take-back programmes, and resale incentives have claimed to achieve positive outcomes. For instance, Patagonia's Worn Wear has recycled and repaired thousands of garments every year, keeping them out of landfills, while H&M's garment collection programme has returned over 37,000 tonnes of post-consumer textiles for recycling or reuse.

Additionally, the industry's reliance on a vast array of chemicals presents a significant concern. A staggering 43 million tonnes of chemicals are used yearly in textile processing, posing a risk to both environmental and human health throughout the product lifecycle and at disposal. The combination of these factors, from excessive resource consumption and greenhouse gas emissions to persistent chemical and microplastic pollution, demonstrates the fundamental failure of the linear economy to create a sustainable textile industry, necessitating a paradigm shift towards more circular and regenerative models.

1.6 PATHWAYS TO ACHIEVING CIRCULARITY IN TEXTILES

1.6.1 Textile Product Design for Circularity

Designing a product to facilitate circularity is a key element of CE (see Figure 1.5). Products are to be designed for durability as well as to facilitate recycling and biodegradation to avoid leaking to the environment. However, these are often conflicting requirements. For example, strong fibres like polyester enhance durability of a textile product but such fibres are derived from fossil fuels and are not degradable. Polyester now accounts for around 57% of global fibre production and over 71 million tonnes are produced annually (8), contributing significantly to the 1.2 billion tonnes of CO_2 generated by the textile industry every year. Adding finishing treatments to reduce abrasion failure or water-repellent finishes may increase durability but such treatments create problems for chemical recycling. For instance, water-repellent perfluorochemicals (PFCs) are widely used in performance textiles, yet they persist in the environment and complicate recycling processes.

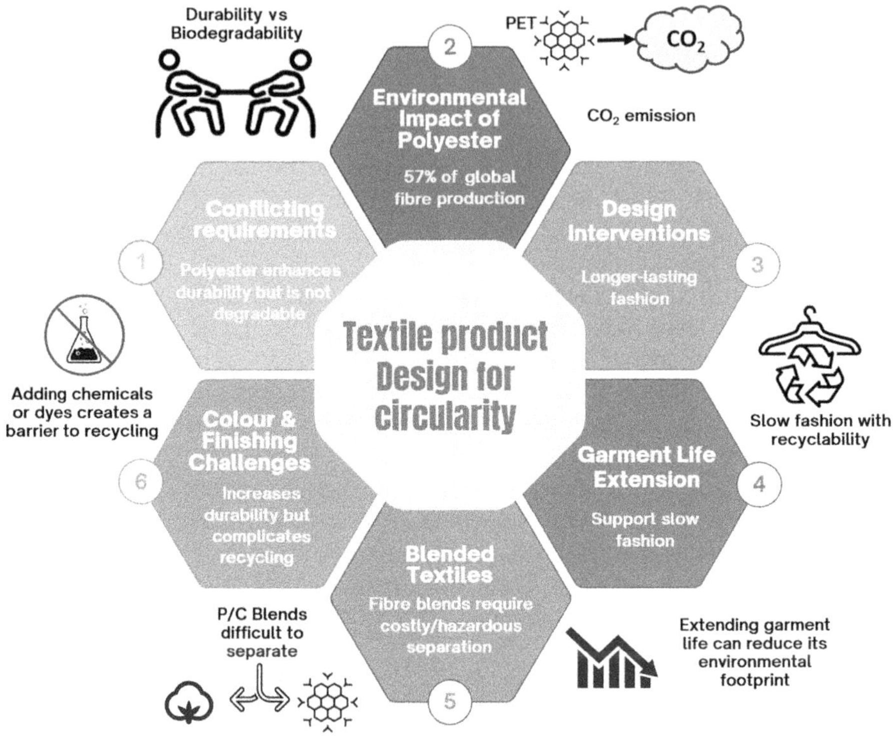

FIGURE 1.5 Textile product design for circularity.

Similarly, colour fastness to washing or light is required for longer use of a product, but such colours are difficult to strip off during recycling. The use of reactive dyes in cotton, for example, poses challenges for chemical recyclers, with colour removal processes adding cost and chemical load. Blend separation requires application of hazardous substances that significantly reduces the positive impacts of recycling and may even create more harm.

Design for circularity also involves designs that sustain long-lasting fashion trends and consumer preferences and not designing to encourage fast fashion. For example, currently available elastomeric fibres are non-biodegradable and stand in the way of recycling. The global elastane (spandex) market produces around 1 million tonnes annually, yet these fibres cannot currently be biodegraded or recycled efficiently, resulting in persistent waste streams. Hence designers need to have a strong understanding of the material properties, such as mechanical and chemical properties, light resistance, and associated durability along with biodegradability and recyclability.

1.6.2 Sustainable Raw Materials for Textile Production

One of the key challenges of CE is the availability of sustainable textile materials. The key raw material for any textile product is fibre. Textile fibres are either natural fibres or man-made fibres. Natural fibres are either plant-based natural fibres such as cotton, lignocellulose fibres like jute or hemp, or animal fibres such as wool or silk. While these fibres are renewable and biodegradable, their production requires intense use of land and water which are limited and compete with more demanding food production. Cotton cultivation also uses about 6% of the world's pesticides and 16% of insecticides, despite being grown on just 2.5% of arable land (11). Hence production of natural fibres over the decades has remained static or declining.

To meet the demand from the textile sector, the production of man-made fibres has been increasing rapidly and now represents more than 75% of fibre produced. In 2021, global fibre production exceeded 113 million tonnes, of which synthetic fibres alone contributed over 63 million tonnes. Made-made fibres are manufactured from synthetic polymers or from natural biopolymers. While biopolymers come from regenerated sources like wood, bacterial cellulose, or recombinant proteins, most manufactured fibres are spun from non-biodegradable polymers such as polyester, nylon, or polypropylene. Polyester dominates this category, accounting for nearly 54–60% of global fibre use, with production projected to reach 92 million tonnes by 2030. Nylon contributes about 5%, while polypropylene accounts for around 2% of total fibre output. These polymers are synthetized from fossil fuel-based raw materials, predominantly produced by the petrochemical sector.

Development of fibres from degradable synthetic polymers such as polylactic acid or other biodegradable polyesters is taking place, but they are a very small part of the industry. Currently, bio-based and biodegradable fibres represent less than 1% of the total fibre market. Their low uptake in the industry is attributed to high cost primarily due to economy of scale and properties inferior to commonly used fibres such as polyethylene terephthalate (PET). PET is non-biodegradable and represents nearly 60% of total fibre used by the sector, growing its share each year. The reliance on PET contributes significantly to environmental challenges, as PET-based textiles release microfibres—estimated at 500,000 tonnes annually—into oceans through washing, highlighting the urgency of developing sustainable alternatives.

The CE means there should be greater use of man-made fibres based on renewable and biodegradable biopolymers. Such biopolymers can be produced using food or agro-industry waste without relying much on trees, land, and water. For example, using agricultural residues like corn husks, sugarcane bagasse, or rice straw can produce cellulose- or starch-based fibres, reducing the competition for arable land. Globally, around 1.3 billion tonnes of food and agro-industrial waste are generated annually, representing a huge potential feedstock for sustainable fibre production.

Other types of sustainable future fibres are those produced using synthetic biology such as fibre-forming recombinant protein. One example is recombinant spider silk fibre. There are commercial fibres produced from such recombinant protein by innovative industries such as Spiber in Japan. It is essential that for CE, the primary materials used to produce the fibres are renewable and not fossil fuel-based like standard PET. To be truly sustainable, the fibres should be produced from regenerative biopolymers as well as those that biodegrade at the end of life and during regular life span should be recyclable.

Innovations like CELYS™ fibres are changing this scenario. CELYS fibres are like PET but have hydrolysable linkages in the polymer backbones, facilitating hydroxylation, thus accelerating the biodegradation rate. Studies show that CELYS fibres can degrade in industrial composting conditions within 6–12 months, compared to conventional PET which takes over 200 years. This biodegradation removes the concerns of microfibre contamination, which contributes an estimated 500,000 tonnes of microplastics annually to oceans from textile washing (12).

1.6.3 Sustainable Production and Regulatory Interventions

The entire value chain of the textile sector has significant impact on use of energy, water, chemicals, and carbon emissions. In the textile CE space, the production sector has been proactive in sustainable practices due to cost imperatives from energy and effluent treatments, pushed by regulatory mechanisms on use of harmful chemicals, and disposal of waste. This is important as current textile production has significant wastes in all production processes such as spinning, fabric production, and garment making. For example, fabric offcuts in garment production can account for up to 15% of the total fabric used, while spinning waste in cotton processing may reach 5–10% of raw material weight (13). Specifically, there have been developments in knitting technology without needing cutting and sewing. For example, a body, arms, and collar are produced simultaneously in a single system without being produced separately, reducing material waste by up to 20–25% compared to conventional cut-and-sew methods.

In the dyeing sector, there are examples of technologies that minimize water use including development of waterless dyeing technologies such as use of supercritical CO_2 instead of water to dye fabrics. Supercritical CO_2 dyeing can eliminate up to 90% of water use in conventional dyeing processes, with energy savings of around 25–30% (14). Some water is however needed in fabric pre-treatments or post-dyeing though this significantly reduces overall water consumption. Other sustainable dyeing technologies are being developed. Natural dyes, both plant and microbial dyes, have also received significant interest as a sustainable alternative to synthetic colour. The global market for natural dyes is projected to grow at a CAGR of 6.2% from 2024 to 2030, reflecting increasing consumer demand for sustainable textiles.

The textile industry uses around 2,450 different chemicals, 1,150 of which are classified as hazardous (15). Regulation plays an important role in identifying producers' roles and responsibilities. The Zero Discharge of Hazardous Chemicals (ZDHC) Foundation is raising awareness and environmental standards in the textile sector (16). It has a ZDHC Manufacturing Restricted Substances List (ZDHC MRSL) that prohibits intentional use of listed chemicals in textile processing. Compliance with ZDHC MRSL standards can reduce toxic chemical discharge by up to 70% in wastewater, while also improving worker safety and reducing costs of effluent treatment. Regulations play a significant role in sustainable manufacturing. High costs on storage, discharge, personal protection, etc. of less sustainable chemicals can stimulate innovations to avoid the challenges of using such chemicals.

Extended producer responsibility (EPR) is a key part of CE and countries such as France and Germany have implemented EPR for textiles. Producers are responsible for the collection, sorting, and recirculation of their products not only prior to sale but also when these are discarded by end users. In France, the EPR scheme for textiles, implemented in 2022, requires producers to ensure that at least 70% of collected textile waste is reused or recycled, shifting an estimated €500 million per year in waste management costs from taxpayers to producers. For example, it shifts the costs of collection of textile waste from taxpayer-funded government to the producers of textiles. The suppliers, retailers, and online platforms will be bound by EPR on textiles.

The extra burden of cost of managing waste is likely to encourage producers to reduce waste and increase circularity by designing products that support CE. The Ellen MacArthur Foundation estimates that scaling up EPR and associated infrastructure could divert 85% of post-consumer textile waste from landfills, creating a market of €9–12 billion annually for reuse and recycling industries. EPR is also needed to attract capital investments in the infrastructure for reuse and recycle at scale, enabling the development of sorting, shredding, and fibre-to-fibre recycling facilities. Overall, regulatory frameworks, technological innovations in production and dyeing, and EPR policies are critical drivers to reduce resource consumption, chemical hazards, and waste, paving the way for a truly circular textile sector.

1.6.4 Durable Use of Textile Products

Durability of a textile material depends not only on material properties but also on consumer education towards care of textile products, thereby playing an important role in textile CE. Proper maintenance and handling of garments can significantly extend their lifespan, reducing the need for frequent replacement and lowering the environmental footprint of textile consumption.

Consumer education includes providing clear washing guidelines, recommending appropriate washing temperatures, detergents, and washing machine cycles to reduce energy and water consumption. For instance, washing clothes at 30°C instead of 60°C can reduce energy consumption for washing by approximately 40–50%, while also minimizing wear and tear on fabrics, extending garment life (17). Proper care can also limit microfibre shedding—a major contributor to microplastic pollution in aquatic ecosystems. It is estimated that global washing of synthetic textiles releases between 0.5 and 1.0 million tonnes of microfibres annually into oceans, representing roughly 35% of total microplastic pollution (18). Reducing washing frequency, using laundry bags, and adopting washing machines with fibre filters can reduce microfibre emissions by up to 80% (19).

Air drying is another important practice in promoting durability and sustainability. Tumble drying can degrade fibres faster due to high temperatures and mechanical stress, increasing fibre breakage and reducing garment lifespan. Studies show that air drying garments instead of machine drying can increase the useful life of clothing by 25–30% over five years. Proper storage, including folding or hanging garments in appropriate conditions to prevent deformation, colour fading, or pest damage, also prolongs garment life. For example, storing woollen products in moth-proof and humidity-controlled conditions reduces fibre loss by over 15% annually.

Repair of garments plays a crucial role in enabling circularity. According to WRAP (UK), repairing garments instead of discarding them can extend their life by 3–5 years, significantly reducing textile waste—up to 20 kg per person per year in high-consumption countries. Easy repair strategies include sewing small tears, replacing buttons or zippers, reinforcing high-stress areas, and patching worn sections. Designs that facilitate repair—like modular clothing, detachable components, and use of standardized materials—make repair more accessible and cost-effective.

Retailers and brands can play a major role in consumer education and facilitating repair facilities. Patagonia provides repair services at over 75 service centres worldwide, repairing thousands of garments each year, helping reduce consumption and landfill contributions. Incentivizing returns of used garments to stores via credit vouchers encourages consumers to participate in circular systems, with studies showing a 25–35% higher return rate for programmes that provide tangible incentives.

Retailers can also support renting and resale platforms, which are gaining popularity globally. The global clothing rental market is projected to reach $2.23 billion by 2028, growing at a CAGR of 9.5%, while the global resale market is expected to exceed $50 billion by 2028, up from $33 billion in 2022 (20). These platforms allow multiple users to benefit from a single product over its lifespan, drastically reducing the demand for new textile production and decreasing associated energy, water, and chemical use. Brands such as Rent the Runway, Le Tote, and The Real Real are leveraging these platforms to promote circular consumption.

Digital tools also support consumer education. Apps and QR codes embedded in clothing can provide care instructions, repair tutorials, and recycling guidance. Studies indicate that 68% of consumers are more likely to follow proper garment care instructions when guidance is accessible digitally (21). Furthermore, educating consumers about the environmental impact of fast fashion such as CO_2 emissions, water consumption, and microfibre pollution can drive behavioural change towards more sustainable practices.

In addition to individual consumer behaviour, collaborations between brands, NGOs, and policymakers can amplify impact. Public awareness campaigns in Europe, such as the "Love Your Clothes" initiative, have demonstrated that targeted education can reduce textile disposal by 20–30% per household. Retailers can facilitate community repair workshops, DIY tutorials, and incentives for sustainable laundry practices. By promoting these measures, consumers become active participants in the CE, helping maintain product value, reduce leakage into waste streams, and conserve natural resources.

Overall, consumer education in textile care, combined with retailer support through repair, take-back, and resale programmes, represents a critical lever in the CE. Proper maintenance, responsible washing, air drying, and repair can extend the functional life of textiles, reduce microfibre pollution, lower water and energy use, and promote a culture of reuse and sustainability, thereby aligning consumer behaviour with CE objectives.

1.7 KEY CHALLENGES OF CURRENT SYSTEMS FOR CIRCULARITY

1.7.1 Collection and Sorting of Textiles

The major challenges of the current system that act as a barrier to circularity in textiles are of three categories—collection, sorting, and recycling challenges (see Figure 1.6). Infrastructure for convenient widespread collection of clean post-consumer textiles of

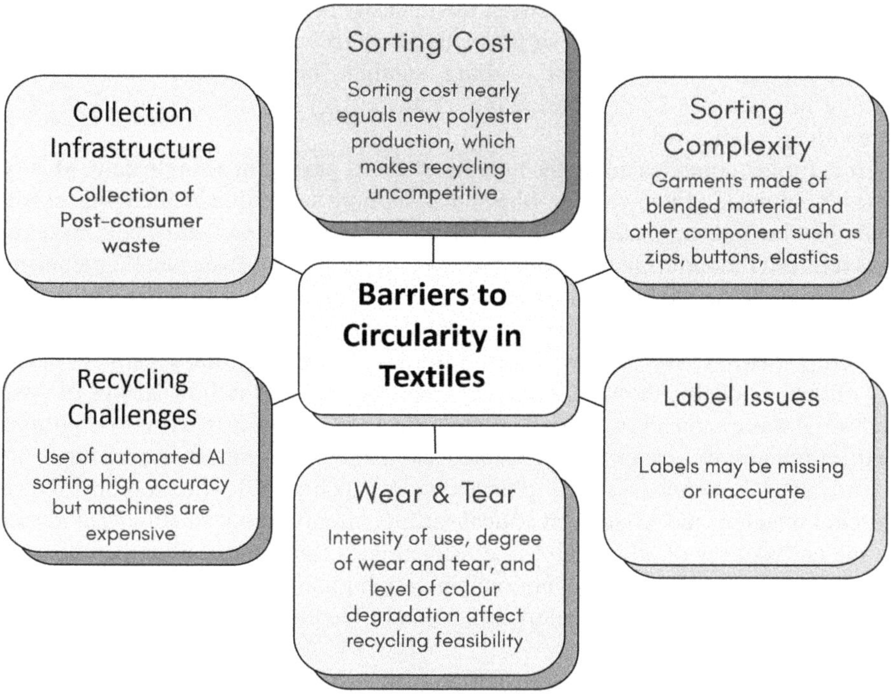

FIGURE 1.6 Barriers to circularity in textiles.

commercially relevant volume is not well established in many countries. Collection systems are slowly evolving in Europe, Australia, and a few other countries. In Europe, for example, only 25–30% of post-consumer textiles are formally collected for reuse or recycling, while in Australia, as per a report from Seamless, the textile product stewardship scheme, the figure is below 10%. In most developing countries, formal collection rates are even lower, often under 5%, with a significant portion of discarded textiles ending up in landfills or informal dumps. This lack of infrastructure presents a critical bottleneck for achieving circularity at scale.

The sorting of collected textiles is still largely done using manual labour which is costly and thus products sorted for reuse cannot compete with low-cost virgin textiles. Manual sorting can be expensive particularly in the developed world where per capita consumption of new items is very high. It makes recycled products less competitive compared to low-cost virgin textiles, such as polyester produced from petrochemical sources. In the case of fibre-to-fibre recycling, sorting of products based on fibre type and colour is critical. Textiles are made of different natural and synthetic polymers and hence sorting is a complex matter.

It is not possible to visually identify fibre composition with a high degree of accuracy. Fibre content labels may be missing or incorrect. Moreover, variation due to intensity of use, level of degradation of colour and fibre play critical roles in recycling of post-consumer textiles. Studies indicate that fibre labels are inaccurate in up to 30% of garments in certain markets, leading to errors in sorting and reducing the quality of recycled outputs. Moreover, the intensity of use, degree of wear and tear, and level of colour degradation significantly affect recycling feasibility. Textiles that have been exposed to repeated washing, sunlight, or chemical treatments often exhibit fibre shortening or crosslinking, reducing recyclability and the mechanical strength of regenerated fibres.

It is further complicated by the number of fibres present in a single item. Mostly, textiles are made of two or more fibres, making fibre separation challenging, costly, and often inefficient. In addition, materials such as zips, buttons, and labels make the problem even more difficult to sort materials for recycling. Even small amounts of metal or elastomeric fibres can contaminate recycling streams, lowering the quality of regenerated material and sometimes rendering a batch unusable. As a result, most recycling today is associated with post-industrial waste, particularly garment industry offcuts. Globally, the textile industry generates over 20 million tonnes of post-industrial waste annually, which is easier to collect and sort compared to post-consumer textiles because it is clean, homogeneous, and free of user-induced degradation.

Advances in automated sorting technologies, such as NIR spectroscopy, hyperspectral imaging, and AI-assisted optical sorting, offer potential solutions but remain in the early stages of adoption. These technologies can achieve fibre identification accuracy, but costs per sorting unit remain very high, making scale-up challenging. Additionally, colour sorting and identification of blended fibres still require further development.

Overall, the lack of widespread post-consumer textile collection infrastructure, coupled with the complexity of sorting garments by fibre type, colour, and component materials, represents a major barrier to circularity in the textile sector. Addressing these challenges through investment in collection systems, automated sorting technologies, and standardization of garment materials is essential to increase fibre-to-fibre recycling, reduce waste leakage, and enable a truly circular textile industry.

1.8 CURRENT STATUS AND FUTURE PROSPECTS OF RECYCLING TECHNOLOGIES

In this section only fibre-to-fibre recycling is discussed when textiles are recycled to produce nearly similar textile products. This section does not cover recycled non-woven products such as insulation for the automotive and construction industries, industrial wipes, and mattress filler since these are examples of repurposing waste with low value addition and considered downcycling of textiles.

1.8.1 Mechanical Recycling

Mechanical recycling is the most adopted technology for fibre-to-fibre recycling. Figure 1.7 provides a summary of the approach used in mechanical recycling. However, it is not truly circular as fibre length is reduced during recovering fibres by pulling from fabric and more than one cycle of recycling is not possible as fibres become too short for a 2nd cycle. Mechanical recycling is limited mostly to post-industrial knitted waste as it is easy to segregate based on fibre and colour. Knitted products are suitable because fibre damage is less aggressive compared to woven

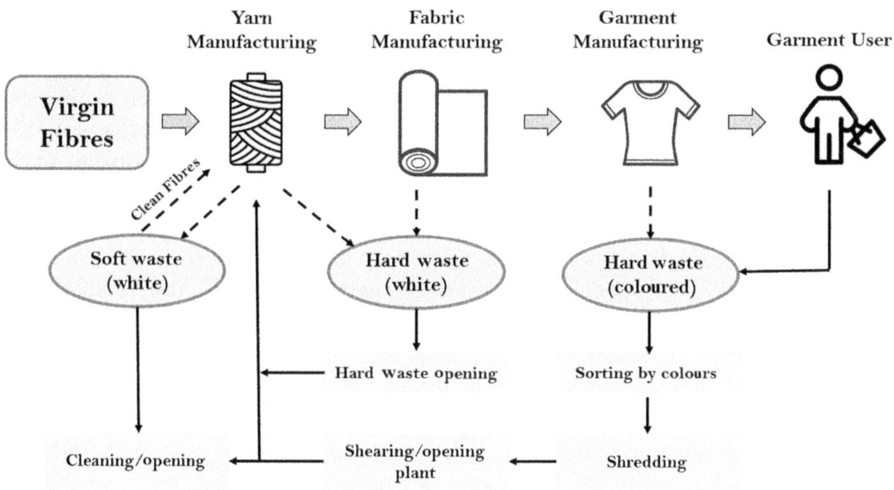

FIGURE 1.7 Mechanical recycling of textile.

textiles during pulling out fibres from garments. However, even for fibres recovered from knitted products, recycled yarn has poor strength and spinning yield is low due to end breakage during spinning. Hence recycled yarns are produced using open-end spinning technology which can handle shorter fibres and often polyester fibres are blended to improve strength and improve spinning performance. At present mechanical recycling is performed mostly on wool and cotton, as it is challenging to open thermoplastic synthetic fibres like polyester as heat generated during shredding softens and melts fibres.

In recent times, there are innovations in equipment for gentle fibre recovery with less reduction in fibre length that allow spinning fine ring-spun yarn from recycled cotton even without the addition of polyester fibre during spinning. This is encouraging as it will facilitate enhanced circularity of textiles. Another key advantage of mechanical recycling that motivates industry is that it can avoid recolouring fibres. Recycling is done following segregation of textiles based on colour and recycled products do not require additional dyeing that can significantly reduce cost to the manufacturers and associated impacts on environment from dyeing.

Globally, mechanical recycling accounts for nearly 82% of the textile recycling market share, highlighting its dominance in practice. The efficiency of recycling is highly dependent on tracking and separation: recent advances in digitalization technologies such as AI-enabled hyperspectral imaging, robotics, and digital twin systems have improved textile waste identification accuracy and throughput, facilitating more precise segregation by fibre type and colour (22). A digital twin system in the textile industry is a virtual replica of physical assets like machines, production lines, or even garments. It collects real-time data from sensors, IoT devices, and production software to simulate, monitor, and optimize operations. These innovations not only improve material recovery but also reduce pre-processing losses. Integration of such digital tools with mechanical recycling is therefore essential to make fibre-to-fibre recycling more scalable and circular.

1.8.2 Chemical Recycling

In the case of chemical recycling, sorted textiles are treated to remove dyes and chemicals and to prepare pulp with adequate molecular weight, followed by dissolving in solvents to prepare a spinning dope. Wet spinning is the most preferred option to spin fibre. The chemical pre-treatments and strong fibre-dissolving solvents make the process challenging from sustainability viewpoints. Recovery and reuse of such solvents are important requirements. The cost and sustainability are two key considerations for successful fibre-to-fibre chemical recycling of textiles.

Chemical recycling technologies are mostly used for cellulosic fibres. Still, post-industrial wastes are predominantly used and both Lyocell and viscose processes are used to spin fibres. Blend separation and separation of elastomers, colours, dyes, and finishing agents, and variation in degradation level of post-consumer textiles pose problems in recycling. For example, Birla Cellulose uses only white and pastel-coloured pure cotton material for Live Reviva, its recycled cellulose fibre brand. The company uses up to 30% pre-consumer and post-consumer waste with wood pulp, contributing to circularity, and has reported a potential to recycle over thousand tonnes of cotton waste annually through such initiatives.

1.8.3 Thermal Recycling

Thermal recycling is adapted to spin fibres from thermoplastic materials. The process requires high energy inputs and the melting process results in reduction in molecular weight and thus fibre-to-fibre recycling is still not adapted at commercial scale. In the case of polyester fibre, PET bottles are used to produce fibre using thermal recycling since molecular weight of PET bottle is higher than fibres and even with thermal degradation molecular weight of melt is adequate for fibre production. However, once recycled into fibre further recycling is not possible due to continuous reduction of molecular weight. For example, studies show that the intrinsic viscosity (IV) of PET recycled from textile fabrics can be as low as 0.58 dL/g, compared to higher values in bottle-grade PET, while its melt flow rate reaches 48.8 g/10 min against 18.6 g/10 min for bottle PET and 14.6 g/10 min for virgin PET (23). A study shows that in bottle-to-bottle recycling, the IV of pellets initially at 0.73–0.80 dL/g typically drops by around 9–14% after reprocessing into bottles, reflecting the persistent challenge of thermal degradation across recycling routes (24).

While bottle-grade PET is clean without dyes and other ingredients, textile fibres contain dyes and various other elements such as UV stabilizers, decluttering agents, and other additives. These materials create additional problems during recycling.

There have been developments in extrusion technology where viscosity of melt is increased under vacuum enabling spinning of fibres with good properties. In-line melt-viscosity sensors track intrinsic viscosity and degradation during extrusion, allowing process parameters to be adjusted dynamically including blending with virgin polymers to maintain the viscosity and properties of spun fibres (25). Machine learning models are also applied to correlate feedstock quality and processing conditions with final fibre properties, helping to minimize IV loss and improve spinnability (26). These digital interventions make thermal recycling more energy-efficient, reduce property loss in rPET, and extend the practical feasibility of fibre production from recycled thermoplastics.

1.8.4 Recycling via Depolymerization

All synthetic fibres are spun from polymers. When these fibre polymers are depolymerized into monomers or oligomers then it is possible to repolymerize targeting the adequate molecular weight suitable to spin fibres with desired properties. Depolymerization can be performed by chemical route or enzymatic process. One of the key chemical depolymerization processes of PET polyester is glycolysis in which in the presence of trans-esterification catalysts, mainly metal acetates, ester linkages are broken and replaced with hydroxyl terminals. PET is converted into bis(hydroxyethyl) terephthalate (BHET) and its oligomers. Subsequent polymerization results in PET for fibre spinning. One of the problems is that oligomers are very difficult to remove by economically viable procedure. Recovered BHET is often blended with virgin BHET during polymerization to maintain the quality of final polymer and fibres.

Enzyme-based depolymerization is another route which is ideal from sustainability viewpoints (see Figure 1.8). Enzymes target specific molecular bonds within the polymer backbone. Esterases, for example, can work on polyester fibres to create shorter oligomers and eventually monomers. Degradability depends on the spacing of the ester groups and the crystallinity in fibre (27). It is hard to depolymerize most synthetic fibres using enzyme alone as enzymes often do not act on the crystalline

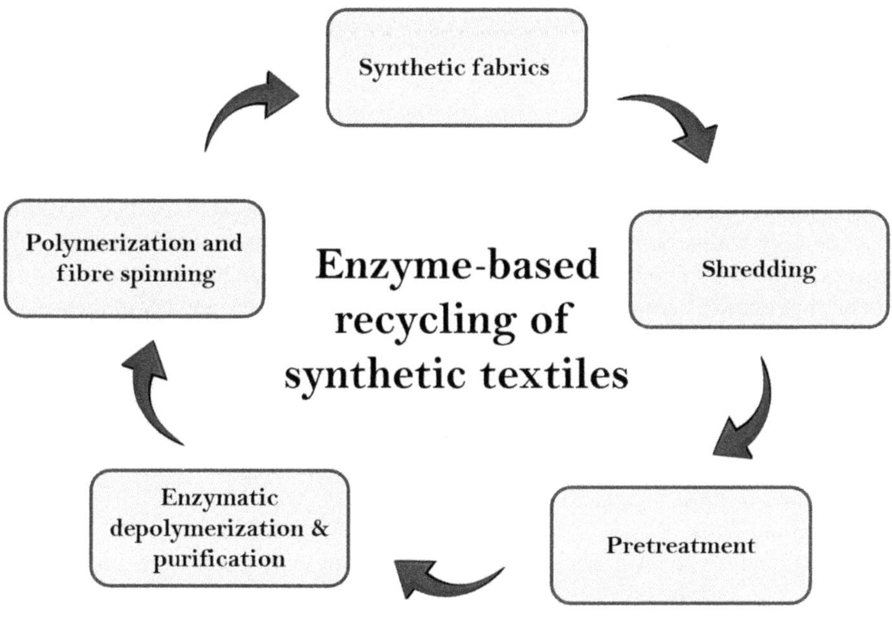

FIGURE 1.8 Enzyme-based recycling of synthetic textiles.

domains. Therefore, pre-treatments are needed to convert fibres to amorphous polymers following which enzymatic depolymerization is adapted to get monomers. Amorphous polymer is produced either by melting and rapidly cooling the melted polymer avoiding the formation of crystals or by chemical route to dissolve the crystals and make an amorphous polymer which can be depolymerized by enzymes. This approach of enzymatic depolymerization is considered more sustainable than chemical approaches such as glycolysis. Developing appropriate enzymes effective for breaking down the specific covalent bonds in the polymer backbone to create oligomers and monomers is the key challenge. Development of synthetic biology, machine learning, and Artificial Intelligence is helping to design and produce specific enzymes for each specific polymer. Growth of this area has provided significant hope for textile CE where repeated polymerization and depolymerization may enable true circularity. These innovations are gradually being transferred from laboratory to industrial scale. Examples include German company Carbios for PET recycling and Australian Samsara Eco in developing recycled Nylon 66. In this approach, following depolymerization, monomers are purified and residual dyes and other elements are removed prior to polymerization so that the recycled polymer fibre is ready for the next cycle of textile production with desired properties.

1.8.5 Textile Recycled Standards

The Recycled Claim Standard (RCS) and Global Recycled Standard (GRS) are examples of currently used standards for textile recycling (28). The standards set the criteria for third-party certification of recycled materials. In the certification system a professional third-party certification body audits each stage in the supply chain. The process ensures that the recycled content is maintained from feedstock to final product. For instance, over 3,000 companies worldwide currently participate in GRS or RCS certification programmes, covering more than 100,000 tonnes of recycled textiles annually. The process ensures that the recycled content is maintained throughout, preventing greenwashing and guaranteeing traceability.

GRS includes a minimum of 50% recycled content and additional social and environmental requirements related to processing and chemical use. Certified facilities must comply with strict limits on heavy metals, azo dyes, and other hazardous substances, with compliance monitored through regular water, air, and soil testing. Social requirements cover worker health and safety, fair wages, and compliance with labour regulations. RCS and GRS certifications also enable brands to communicate verified claims to consumers, supporting market trust and enabling CE goals. For example, textiles certified under GRS have contributed to diverting over 20,000 tonnes of post-consumer and post-industrial waste from landfills annually, while promoting responsible chemical management in production (29).

Adoption of these standards is expanding globally, particularly in Europe, North America, and parts of Asia, reflecting growing demand for recycled and traceable textile products. The RCS and GRS frameworks are increasingly being integrated into corporate sustainability strategies, with many brands pledging to achieve 100% recycled content in selected product lines by 2030.

1.9 SUMMARY

The textile sector, one of the world's most resource-intensive industries, is undergoing a transformative shift driven by the principles of CE. Traditional practices consume vast amounts of water, energy, and chemicals. Circularity in textiles addresses these challenges by promoting the efficient use of materials, extending product lifespans, reducing waste, and minimizing environmental impact.

A key focus of CE is the design and use of sustainable fibres. Natural fibres, though renewable and biodegradable, face constraints such as high water and land requirements, limiting their scalability. Man-made fibres derived from fossil fuels, such as polyester, now account for more than 60% of global fibre production, while biodegradable and regenerative fibres, including PLA-based fibres or recombinant proteins like spider silk, remain a small fraction of the market. Circularity emphasizes the adoption of fibres that are renewable, recyclable, and biodegradable, thereby reducing dependence on non-renewable feedstocks and mitigating microfibre pollution.

Circular design principles underpin the extension of textile lifespans. Durability, repairability, modular design, and zero-waste pattern-making are critical strategies to reduce material waste, minimize environmental leakage, and enhance recyclability. Implementing these practices could reduce greenhouse gas emissions by up to 40%, conserve billions of cubic metres of water, and save millions of tonnes of raw materials annually. Consumer behaviour is equally pivotal; proper garment care, washing at lower temperatures, air drying, and facilitating repair or resale can significantly extend product life and reduce environmental impact. Retailers and brands play a central role by providing repair services, take-back programmes, resale and rental platforms, and education initiatives to encourage responsible consumption.

Production processes are evolving to support CE, with innovations such as 3D knitting, waterless dyeing using supercritical CO_2, natural and microbial dyes, and reduction of hazardous chemicals through ZDHC guidelines. EPR policies in countries like France and Germany shift the responsibility for post-consumer textile collection and recycling to producers, incentivizing waste reduction and circular product design.

Promoting natural-fibre farming through government-led initiatives, farmland protection, and contract farming not only addresses the economic challenges of cotton cultivation but also strengthens India's position in the global circular textile economy. By increasing cotton supply, reducing costs, and stabilizing farmer incomes, India can shift its textile sector towards a sustainable trajectory while simultaneously advancing multiple SDG targets.

Overall, CE in textiles integrates sustainable raw materials, innovative design, responsible production, consumer education, and systemic policies to create a regenerative, resource-efficient, and environmentally responsible sector. By closing material loops, reducing waste, and fostering durability and reuse, the textile industry can transition towards a resilient future that balances economic growth with ecological sustainability. Circularity is no longer optional—it is essential for mitigating environmental impact and ensuring the long-term viability of the textile sector.

REFERENCES

1. Ghemawat P, Nueno JL. ZARA: Fast fashion [Internet]. Harvard Business School Publishing, 2006. Available from: https://didierdiaz.com/wp-content/uploads/2019/10/Zara-fast-fashion-Case-study-HVR.pdf
2. BeallA.Whyclothesaresohardtorecycle[Internet].*BBC*[cited2020July13].Availablefrom: https://www.bbc.com/future/article/20200710-why-clothes-are-so-hard-to-recycle
3. Browne MA, Crump P, Niven SJ, Teuten E, Tonkin A, Galloway T, et al. Accumulation of microplastic on shorelines worldwide: Sources and sinks. *Environmental Science & Technology.* 2011 Nov 1;45(21):9175–9.
4. UN Environmental Programme. Fashion's tiny hidden secret [Internet]. 2019. Available from: https://www.unep.org/news-and-stories/story/fashions-tiny-hidden-secret#:~:text=About%2060%20per%20cent%20of,and%20broken%20down%20digestive%20systems
5. Boucher J, Friot D. Primary microplastics in the oceans: A global evaluation of sources [Internet]. IUCN International Union for Conservation of Nature, 2017 [cited 2025 Sept 22]. Available from: https://portals.iucn.org/library/node/46622
6. Nimkar U. Sustainable chemistry: A solution to the textile industry in a developing world. *Current Opinion in Green and Sustainable Chemistry.* 2018 Feb;9:13–7.
7. Reinhart J. Thredup resale report 2023 [Internet]. 2023 Apr. Available from: https://cf-assets-tup.thredup.com/resale_report/2023/thredUP_2023_Resale_Report_FINAL.pdf
8. Textile Exchange. Materials market report [Internet]. 2024 Sept, p. 1–76. Available from: https://textileexchange.org/app/uploads/2024/09/Materials-Market-Report-2024.pdf
9. Shamsuzzaman Md, Hossain I, Saha T, Roy A, Das D, Ahmed MdT, et al. Waste management in textile industry. In: Rahman MdM, Mashud M, Rahman MdM, editors. *Advanced Technology in Textiles* [Internet]. Singapore: Springer Nature Singapore; 2023 [cited 2025 Oct 1]. p. 279–99. (Textile Science and Clothing Technology). Available from: https://link.springer.com/10.1007/978-981-99-2142-3_10
10. One T-Shirt = 2,700 liters of water?! Yep, you read that right. [Internet]. 2025. Available from: https://www.caelusrome.com/blogs/blog-posts/how-much-water-does-one-tshirt-take?srsltid=AfmBOop-jbXQAJcgkBm4gFCATvd3IEWvGxPJd4qNIKjCqA0fuO1XLXWk
11. Karaş E. Sustainable and effective management strategies in cotton cultivation. In: YI Abdurakhmonov, editor. *Cotton* [Internet]. Intech Open; 2022 [cited 2025 Oct 4]. Available from: https://www.intechopen.com/chapters/81421
12. Chatziparaskeva G, Papamichael I, Zorpas AA. Microplastics in the coastal environment of Mediterranean and the impact on sustainability level. *Sustainable Chemistry and Pharmacy.* 2022 Oct;29:100768.
13. Khairul Akter MMd, Haq UN, Islam MdM, Uddin MA. Textile-apparel manufacturing and material waste management in the circular economy: A conceptual model to achieve sustainable development goal (SDG) 12 for Bangladesh. *Cleaner Environmental Systems.* 2022 Mar;4:100070.
14. Zheng H, Zhang J, Yan J, Zheng L. An industrial scale multiple supercritical carbon dioxide apparatus and its eco-friendly dyeing production. *Journal of CO_2 Utilization.* 2016 Dec;16:272–81.
15. Swedish Chemicals Agency. Chemicals in textiles – Risks to human health and the environment. Report No. 6/14. Kemikalieinspektionen. 2014. Stockholm, *Sweden, p.* 142
16. CEO Water Mandate. Zero discharge of hazardous chemicals (ZDHC) programme [Internet]. [cited 2023 Apr 14]. Available from: https://ceowatermandate.org/resources/zero-discharge-of-hazardous-chemicals-zdhc-programme-2018/

17. Pakula C, Stamminger R. Electricity and water consumption for laundry washing by washing machine worldwide. *Energy Efficiency.* 2010 Nov;3(4):365–82.
18. Ellen MacArthur Foundation. A new textiles economy: Redesigning fashion's future. [Internet]. [cited 2020 Aug 1]. Available from: https://www.ellenmacarthurfoundation.org/assets/downloads/publications/A-New-Textiles-Economy_Full-Report.pdf
19. Abourich M, Bellamy A, Ajji A, Claveau-Mallet D. Washing machine filters to mitigate microplastics release: Citizen science study to estimate microfibers capture potential and assess their social acceptability. *Environmental Challenges.* 2024 Aug;16:100984.
20. Grand View Research. Online clothing rental market [Internet]. p. 80. Report Year: 2022 GVR-3-68038-089-7. Available from: https://www.grandviewresearch.com/industry-analysis/online-clothing-rental-market
21. Cruz EF, Rosado Da Cruz AM. Digital solutions for engaging end-consumers in the circular economy of the textile and clothing value chain—A systematic review. *Cleaner and Responsible Consumption.* 2023 Dec;11:100138.
22. Spyridis Y, Argyriou V, Sarigiannidis A, Radoglou P, Sarigiannidis P. Autonomous AI-enabled industrial sorting pipeline for advanced textile recycling [Internet]. arXiv; 2024 [cited 2025 Sept 26]. Available from: https://arxiv.org/abs/2405.10696
23. Wu WJ, Sun XL, Chen Q, Qian Q. Recycled poly(ethylene terephthalate) from waste textiles with improved thermal and rheological properties by chain extension. *Polymers.* 2022 Jan 27;14(3):510.
24. Alvarado Chacon F, Brouwer MT, Thoden Van Velzen EU. Effect of recycled content and rPET quality on the properties of PET bottles, part I: Optical and mechanical properties. *Packaging Technology and Science.* 2020 Sept;33(9):347–57.
25. Muringayil Joseph T, Azat S, Ahmadi Z, Moini Jazani O, Esmaeili A, Kianfar E, et al. Polyethylene terephthalate (PET) recycling: A review. *Case Studies in Chemical and Environmental Engineering.* 2024 June;9:100673.
26. Ciano MP, Peron M, Panza L, Pozzi R. Industry 4.0 technologies in support of circular economy: A 10R-based integration framework. *Computers & Industrial Engineering.* 2025 Mar;201:110867.
27. Bai Z, Liu Y, Su T, Wang Z. Effect of hydroxyl monomers on the enzymatic degradation of poly(ethylene succinate), poly(butylene succinate), and poly(hexylene succinate). *Polymers.* 2018 Jan 18;10(1):90.
28. Lin HJ, Ma H Wen. Analysis of green certification standards related to recycled materials involving textiles based on life cycle thinking. *Sustainable Production and Consumption.* 2023 Oct;41:107–20.
29. Arafat Y, Uddin AJ. Recycled fibers from pre- and post-consumer textile waste as blend constituents in manufacturing 100% cotton yarns in ring spinning: A sustainable and eco-friendly approach. *Heliyon.* 2022 Nov;8(11):e11275.

2 An Overview of Textile Processes and Waste Resources

2.1 INTRODUCTION

The global textile industry is nearly a one-trillion-dollar industry and continues to grow. Waste analytics refers to data on waste generation, impacts on the environment, and waste audits. Waste typically refers to materials that are no longer usable once a process is complete or after the primary function of a product has been fulfilled. In the context of textiles, waste refers to materials that become unusable or lose their value either during the manufacturing process or at the end of the product's life cycle. Worldwide, approximately 92 million tons (MT) of textile waste are generated each year. According to data from the Central Pollution Control Board, India generates around 6.23 MT of textile waste annually (1). Globally, 75% of textile waste is dumped into the ground, the remaining 25% being repurposed, and a mere fraction—less than 1%—is converted into value-added products (2). Survey results indicate that 17% of respondents view disposal as the most convenient and fastest method for managing unwanted textiles, with 10% citing convenience and 7% highlighting the speed of disposal (3). This preference underscores the challenge of increasing recycling and reuse rates, as many people prioritize ease and quickness over sustainable practices. Wastage poses a significant threat to any textile industry, leading to value loss, and also presents disposal constraints to the environment.

Textile production has a significant effect on the environment. For instance, producing a single T-shirt and a pair of cotton pants requires 713 gallons of water—an amount that could sustain a person's drinking needs for two years (4,5). The production process also generates substantial greenhouse gas emissions: cotton harvesting emits approximately 4.7 kg carbon di oxide (CO_2) emissions per kilogram of fibre, plastic-based fibres produce around 11.9 kg CO_2 emissions per kilogram, and the manufacturing of yarn and fabric emits about 9.6 kg CO_2 emissions per kilogram (6). In response to these environmental challenges, a recent report by Textile Exchange recommends a 30% reduction in CO_2 emissions from textile manufacturing by 2030, emphasizing the significance of a circular economy to mitigate these impacts and address the urgent need for climate action (2).

Moreover, recycling second-hand clothes can significantly reduce environmental harm by cutting 53% from greenhouse gas emissions, 45% pollution from chemical processing, and enriching water with nutrients by 95% (7). Considering the amount

DOI: 10.1201/9781003503835-2

of waste generated by textile processes and its impact on the environment, it is of paramount importance to utilize textile waste. The solution for this would be recycling and adding value to waste, as evidenced by the identification of 3,296 papers and the involvement of 10,451 authors in textile waste research (8). In 2022, India's textile recycling market was valued at US$308.7 million and is projected to grow to US$375 million by 2028 (9).

2.2 SOURCES OF TEXTILE WASTE

The global population, which was 7.5 billion in 2006, increased to 8.1 billion by 2023, and is estimated to reach 8.5 billion by 2030. This population growth is driving a continuous rise in global demand for textile products, a trend that shows no signs of slowing down (10). In 1975, global textile fibre production was around 24 million metric tons. By 2022, this figure had surged nearly fivefold, surpassing 113.8 million metric tons. Forecasts suggest that by 2030, worldwide fibre output could reach 149 million metric tons, representing a substantial rise compared to 2020 levels (11). This surge in production is in response to the ever-growing need for textiles driven by population growth and rising consumer demand.

In 2023, the global textile industry was estimated at approximately $1,837 billion. Forecasts predict that this figure will grow to around $3,047 billion by 2030, representing a steady compound annual growth rate (CAGR) of nearly 7.4% during the forecast period (12). This growth is fuelled by various factors including increasing disposable incomes, rapid urbanization, and the rising trend of fast fashion which demands quick and large-scale production of textile goods.

Textile waste is generated throughout multiple stages of the manufacturing process, including fibre production, spinning, weaving, dyeing, finishing, garment production, and even at the consumer level. A core issue within the fast fashion and textile sectors is the significant waste produced at each of these stages. The fashion industry's supply chains are highly fragmented and extensive, often leaving retailers disconnected from the processes involved in sourcing raw materials and producing finished goods.

In the textile and clothing industry, waste can be broadly categorized into three main types: production waste, pre-consumer or post-industrial waste, and post-consumer waste. Production waste arises during the manufacturing processes. This includes scraps and by-products generated at various stages of textile production, such as ginning, spinning, weaving, dyeing, and garment making. These wastes can consist of seedy fibre, fibrous waste of sliver, roving, leftover fabric, trimmings, and defective products that never make it past the factory floor. Pre-consumer waste, also known as post-industrial waste, originates from products that are damaged or deemed unsellable before they reach the consumer. This can occur both in stores and during the production phase. Examples include garments that are flawed or excess inventory that cannot be sold. Post-consumer waste is the final category and refers to items that have been used and discarded by consumers. This includes worn-out or outdated clothing and textiles that individuals throw away. Post-consumer waste represents a significant challenge for the industry due to the vast quantities of textiles discarded by consumers each year.

Addressing these types of waste is crucial for improving sustainability in the textile industry. By reducing production waste, reusing soft waste, repurposing pre-consumer waste, and recycling post-consumer waste, the industry can significantly lower its environmental impact. Strategies such as enhancing manufacturing efficiency and work practices, adopting circular economy principles, and encouraging consumer recycling can play pivotal roles in managing textile waste effectively (13).

2.3 CATEGORIZATION OF TEXTILE WASTE

Textile waste can be categorized based on various parameters, including:

- Classification based on the origin of the waste: pre-consumer waste, post-consumer waste, industrial waste, and agricultural waste.
- Classification based on reusability: hard waste and soft waste.
- Classification based on the industry: spinning, weaving, knitting, processing, and garments.
- Classification based on defects: package defects, fabric defects, and garment defects.
- Classification based on the safety aspect of the waste: hazardous waste and non-hazardous waste.
- Classification based on the physical properties of the waste: compostable and recyclable waste.

Amongst these classifications, the most commonly used categorization for classifying textile waste is based on the origin and industry of the textile waste.

2.3.1 Waste Categorization Based on Origin

Textile waste is primarily categorized into three distinct types based on its origin: pre-consumer waste, post-consumer waste, and industrial waste.

2.3.1.1 Pre-Consumer Waste

In the textile industry, pre-consumer or post-industrial waste is generated unintentionally or due to manufacturing and processing faults, making materials unsuitable for their original intended purpose. A study carried out at OCAD University in Ontario, Canada, found that nearly 35% of materials used throughout the supply chain are discarded as waste prior to the garment or product reaching the end consumer (14). In addition, nearly 47% of the fibres introduced into the fashion value chain are discarded as waste during different phases of production, ranging from fibre conversion to yarn, fabric manufacturing, and ultimately garment creation.

These insights underscore the considerable inefficiencies within the textile manufacturing process. Research indicates that merely around 65% of raw input materials are transformed into finished clothing and related products for consumers. Consequently, approximately 35% is lost as waste during production and pre-consumer phases. This waste comprises fabric offcuts, defective items, and surplus stock that never reach retail shelves.

FIGURE 2.1 Pre-consumer fabric wastes.

Pre-consumer waste in the textile industry primarily occurs for several reasons (15). First, manufacturers often over-order materials by 3–10% to ensure they have enough for production. Luxury brands sometimes even destroy overstock to protect their brand identity and exclusive designs. Second, inconsistencies in fabric production, such as weaving defects, printing mistakes, and colour variations, contribute significantly to waste (see Figure 2.1). Third, cutting residues arise when fabrics are trimmed into specific forms, resulting in 8–15% of the fabric, known as textile scraps and roll ends, being discarded.

Another form of pre-consumer waste stems from cutting and sewing errors caused by human negligence, including tiny holes, stains, and oil marks. For instance, grease stains on pure white fabric during the weaving process render it unsuitable for its intended use. Additionally, unsold and damaged products at the retail level contribute to this type of waste. Repairing garments can be a quick and cost-effective solution with significant benefits for manufacturers and brands. In small manufacturing units, fixing just 3% of defective garments can save about USD 1 million annually, while a brand-wide enterprise to repair the same percentage could save up to USD 1 billion. With total waste comprising 25.6% of defective fabric, defective garments, and cutting residue, repairing just 10% of this waste could result in savings of several billion dollars (16). Implementing repair strategies not only reduces waste but also offers substantial financial benefits, highlighting the significance of sustainable goals in the textile industry.

2.3.1.2 Post-Consumer Waste

Textile waste refers to items that have exceeded their usable lifespan and are no longer needed or suitable for their original purpose, commonly seen in discarded clothing due to wear and tear, colour fading, or fast fashion trends (see Figure 2.2). The clothing industry contributes significantly to this waste stream, driven by the rapid turnover of fashion trends, known as fast fashion (17). Research indicates that worldwide fashion waste reached 92 MT in 2015 and is expected to rise to 148 MT by 2030 (18). In the USA alone, 16 MT of fashion waste were generated in 2015, while only a minimal amount was recycled (15.28%), and a larger portion was either incinerated (19.03%) or sent to landfills (19).

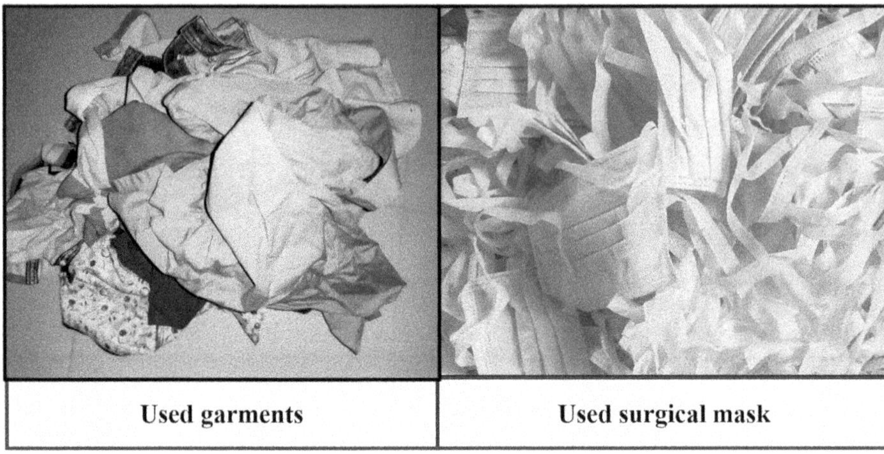

FIGURE 2.2 Post-consumer wastes.

The predominant method of disposing of such waste is through municipal landfills. Data from the Environmental Protection Agency reveals that the United States produces over 17 MT of textile municipal solid waste annually, equating to about 112 pounds per person (20). However, not all waste ends up in landfills; some materials are donated to charities or repurposed based on their condition.

In developing countries like Sri Lanka, study reveals that consumer disposal behaviours are similar to those in developed economies, except for the practices of burying and burning post-consumer textile waste (PCTW) due to local waste management infrastructure. Sri Lanka, like many developing countries, faces challenges such as poor solid waste management combined with inadequate collection and recycling systems for PCTW. The absence of source separation, recycling technologies, and public awareness about the value of PCTW hinders market creation for recycled textiles and the achievement of sustainable consumption and production goals by 2030. To mitigate these issues, it is crucial to explore opportunities for developing new business models for collecting, sorting, and cleaning PCTW, implement educational programmes to raise public awareness about the environmental impact of textile waste, and establish appropriate policy tools for managing PCTW (21).

2.3.1.3 Industrial Textile Waste

This refers to waste generated from textile processes and products used for various industrial and commercial applications instead of conventional clothing and apparel applications. This kind of waste requires separate techniques for recycling or reuse into new products. Examples of industrial textile waste include carpets, masks, wet wipes, surgical drapes, and filtration media (see Figure 2.3).

The COVID-19 pandemic spurred the textile industry to prioritize the development of materials aimed at safeguarding individuals, leading to the production of personal protective equipment (PPE) like protective gowns, surgical shoes, and masks. These items, typically crafted from materials such as polyacrylonitrile (PAN), polypropylene (PP), polyurethane (PU), or their blends, have become essential for human protection during the outbreak (22,23). However, the widespread adoption of

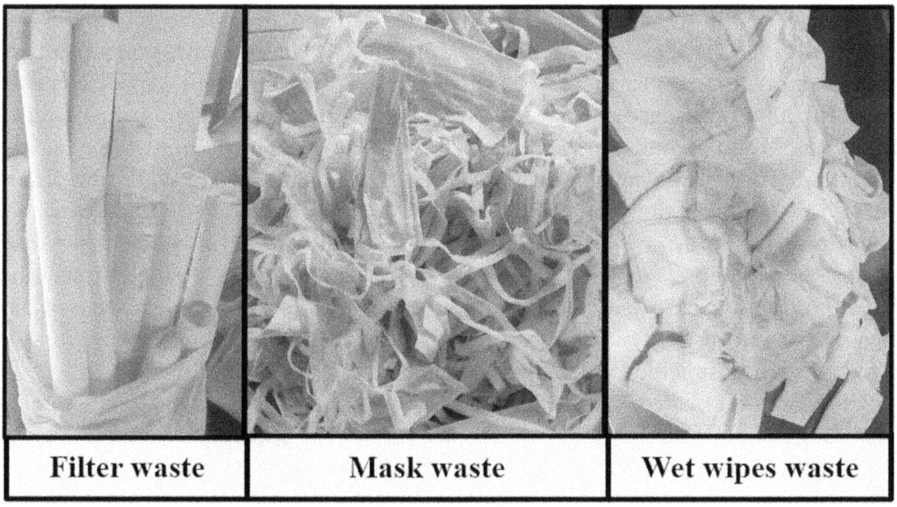

FIGURE 2.3 Industrial textile wastes.

such PPE has resulted in a concerning surge in PPE waste globally, posing environmental complications.

For instance, a study conducted in Peru highlighted the significant impact, with approximately 74.9 tons of plastic waste generated daily from face masks alone, totalling 27,344.7 tons per year (24). In the USA, the utilization of medical masks during the pandemic reached staggering figures, with an estimated 89 million masks used, further exacerbating the issue of PPE waste (25). The disposal of these items, primarily composed of materials like polypropylene and polyethylene, contributes to the proliferation of microplastic pollution in the environment.

Meanwhile, the demand for carpets has remained robust, driven by factors such as interior decoration trends, affordability, and easy replacement options. However, this popularity has led to a substantial volume of carpet waste being generated, posing challenges for landfill management. In California alone, approximately 251 million pounds of carpet were discarded in 2020, accounting for a significant portion of landfill space (26).

Carpets typically consist of plastic and polymer fibres, predominantly PET, nylon, and polypropylene. However, innovative approaches such as recycling waste nylon extracted from discarded carpets offer promising solutions to mitigate environmental impacts. Replacing virgin nylon with recycled nylon derived from carpet waste not only reduces greenhouse gas emissions but also minimizes the depletion of fossil fuels required for nylon production. Studies suggest that this substitution could potentially decrease greenhouse gas emissions by 7.5 metric tons, reduce particulate matter (less than 2 micrometres) by 2 kilotons, and conserve fossil fuel resources by 15 trillion joules (27,28).

2.3.2 Waste Categorization Based on Reusability

In this type of classification, the waste is mainly divided into two categories: soft waste and hard waste.

2.3.2.1 Soft Waste

Soft waste mostly generated during yarn production consists of fibrous materials that retain their structure and can be repurposed within the textile supply chain or converted into other usable products. These waste fibres possess an open structure, facilitating their utilization in the production of lower-grade yarns or as early feeding material for open-end yarn production. Soft waste is mainly produced during different phases of the spinning process, such as in the blow room, carding, draw frame, comber, and speed frame stages. Several factors contribute to soft waste generation during the carding stage, including low-quality feed material, inefficient material handling, high production speeds, improper machine settings, disturbances in air flow in the web delivery zone, poor operational practices, incorrect sliver trumpet selection, and unsuitable environmental conditions (29).

Comber noil is indeed one of the soft wastes used in the open-end spinning process. Comber noil is a by-product of the combing process in cotton yarn production (see Figure 2.4). It consists of short fibres, neps, and impurities removed from the cotton during the combing stage, which is a process designed to remove the shorter fibres and impurities to produce high-quality yarn. Typically, the percentage of comber noil in the fibre mix for open-end spinning ranges from 5% to 15%. However, in some cases, mills may use a higher percentage, up to 25%, depending on the quality of the comber noil and the intended use of the final yarn. The use of comber noil can affect the strength, evenness, and appearance of the yarn, so it is carefully controlled to achieve the desired balance between cost and quality.

FIGURE 2.4 Soft wastes generated from comber machine.

2.3.2.2 Hard Waste

Hard waste is defined as the portion of waste that cannot be reused in its current form, often requiring additional processing or disposal. This waste can include non-fibre materials such as dust, trash particles, and other by-products of the spinning process. The financial impact of hard waste generation in the spinning industry is substantial. This loss accumulates significantly across a large number of spindles, which directly affects the profitability of a spinning mill. Given the scale of operations in such mills, even a slight increase in hard waste can result in considerable financial setbacks.

Several factors contribute to the generation of hard waste in the spinning process, particularly during stages such as cone winding, reeling, open-end spinning, and ring doubling. When yarn ends break, or when feed packages are replaced, there is a higher likelihood of generating hard waste. This is especially true if the process is not managed efficiently, leading to increased waste during unwinding and winding operations. The operational practices of the workforce significantly impact hard waste generation. Inappropriate handling, lack of precision during winding, and operational errors can all contribute to increased waste. For instance, over-piecing, where long ends from the front roller nip or back of the delivery cot result in thick and thin places in the yarn, can cause cuts during winding, leading to more hard waste.

Maintaining high technical standards is essential to minimizing hard waste. Deviations from these standards can result in inefficiencies and increased waste. For example, it is important to routinely check the standard cop diameter, which varies based on the ring diameter, to ensure it remains within the specified range. The back winding coils should generally range from 1.5 to 2.5, with a maximum length of 80 cm, while under winding coils should be maintained between 2 and 3, not exceeding 20 cm in length. Additionally, bobbin hardness is a key parameter and should typically fall between 50° and 55°. Soft bobbins can lead to slough-off, which in turn increases hard waste (see Figure 2.5). Other issues, such as damaged or deformed

FIGURE 2.5 Hard wastes generated from winding machine.

bobbins, long tail ends, and improper winding cuts, also contribute to waste generation. Ensuring that these factors are monitored and controlled is essential for maintaining high production efficiency and minimizing waste (30). In a spinning mill, for profitable operation, operators must ensure that the percentage of hard waste generation does not exceed 1%.

2.3.3 Waste Categorization Based on Industry

2.3.3.1 Ginning Industry

Ginning is the primary mechanical operation in processing raw cotton fibre, involving the separation of cotton fibres from seed bolls and dust particles (see Figure 2.6). During the transition from the harvesting stage to the ginning process, cotton stalks, cotton gin trash (CGT), and cottonseeds are the three by-products generated. Cotton stalks refer to the remaining parts of the plant left in the fields after the cotton bolls are harvested. CGT consists of the waste materials produced during the cleaning stages within the ginning facility. On the other hand, cottonseeds are obtained post-ginning, once the fibres have been removed. These seeds are utilized in cottonseed oil production and for planting new crops, whereas CGT and cotton stalks are considered waste products and must be discarded by the ginners (31).

The quantity of CGT produced varies significantly and is influenced by factors including the harvesting method, ginning process, and the cotton's origin and variety. Global CGT production is substantial, accounting for approximately 32–41% of each cotton bale (32). CGT can frequently be gathered at minimal cost, making its collection economically feasible (33). It is a mixture of different segments of the cotton

FIGURE 2.6 Gin wastes collected in a duct.

plant (fibres, burrs, leaves, and other plant parts), and it is highly difficult to fractionate the constituents due to the entanglement of the fibrous materials. The usage of CGT in the production of composite materials is an area of immense research potential. Studies have revealed that composites made by blending CGT fractions with non-biodegradable polymers like high-density polyethylene demonstrate strong mechanical properties (34). Various studies also reveal the usage of CGT as a feedstock material for the production of bioethanol (35).

2.3.3.2 Spinning Industry

Cotton fibres from bales often contain various types of waste, including foreign particles, dust, seeds, and short fibres. As these fibres are processed through various sections of a spinning mill, various types of waste are produced (see Figure 2.7). Achieving high yarn realization holds significant importance within a spinning mill's operations. Even a minor enhancement of 1% in waste management practices could result in substantial annual savings for a spinning mill comprising about 30,000 spindles producing yarn counts ranging from 20 to 40s (36). This notable cost-saving potential exists because the resale value of waste is much lower than that of cotton until it advances to the yarn stage. Waste generated during the blow room, carding, and combing stages accounts for a major share of total waste—about 80% in cotton spinning mills and 50% in mills handling man-made fibres.

During the cotton processing stages in spinning industries, approximately 8% of waste is generated in the blow room and carding section, while up to 20% noil is produced during combing (37). Additionally, various machines throughout the spinning line produce over 1% cotton fly, which is gathered as filter waste via the humidification system. According to data from the Bangladesh Textile Mills Association (BTMA) in January 2020, around 1.8 MT of raw cotton was brought into Bangladesh

FIGURE 2.7 Wastes generated in different sections of spinning industry.

for spinning yarn in 2019. Assuming that 1% of this cotton's weight is collected as filter waste, it amounts to a total of 18,000 tons of filter waste generated annually (38).

The yarn realization percentage mainly depends on the amount of waste generated during the blow room, carding, and combing processes. Yarn realization of 80 to 82% up to 40s range and 88 to 90% in the 60s to 80s range could be considered reasonable (39). The amount of waste removed in the blow room is largely based on the trash content present in the fibre mix. Waste generated during carding varies based on the type of carding machines used and the level of trash in the lap, while waste in combers is influenced by the fibre length distribution of the mixing used for different combed yarn counts. In synthetic fibre processing, most waste can be reused, with the exception of blow room droppings (approximately 0.5%) and carding flat strips (around 1.5%), since synthetic fibres are generally free of impurities. As a result, synthetic mills typically achieve a yarn realization rate of 97–98%. In comparison, yarn realization for carded cotton yarn ranges between 85–88%, while combed cotton yarn yields about 70–75% (40).

Table 2.1 represents the type of waste generated and its percentage. Wastes like lap ends, sliver fragments, roving tails, and microdust waste are generated at various stages of the spinning process. Conversely, cleaning waste is deliberately removed from cotton to achieve the required yarn standard. This type of waste

TABLE 2.1
Types and Percentage of Spinning Waste Generated

Process	Waste Type	Waste %
Blowroom	Dropping, seed coats, dust, floor sweep, lap waste	3
	microdust waste	3
Carding	Flat strips, licker-in waste, sliver cut waste, filter waste, floor sweep, dropping, vacuum waste, motes, flyer	10
Lap former	Lap waste, sliver waste, vacuum waste	0.5
Comber	noil	12–18
	Mini-lap wastage	0.25
	Sliver wastage	0.25
	Roller wastage	0.25
	Fly dust	0.10
	Sweeping waste	0.20
Draw frame	Filter waste, clearer waste, sliver cut waste	0.5
Speed frame	Roving waste, sliver waste	0.5
Ring frame	Pneumafil	0.20–0.30
	Bonda	0.20–0.30
	Roving waste	0.10–0.20
	Thread waste	<0.10
	Fly dust	0.20
	Sweeping waste	0.20
Winding section	Hard waste	1
Other spinning waste	Fly and dust, directly reusable waste, dirty waste	2

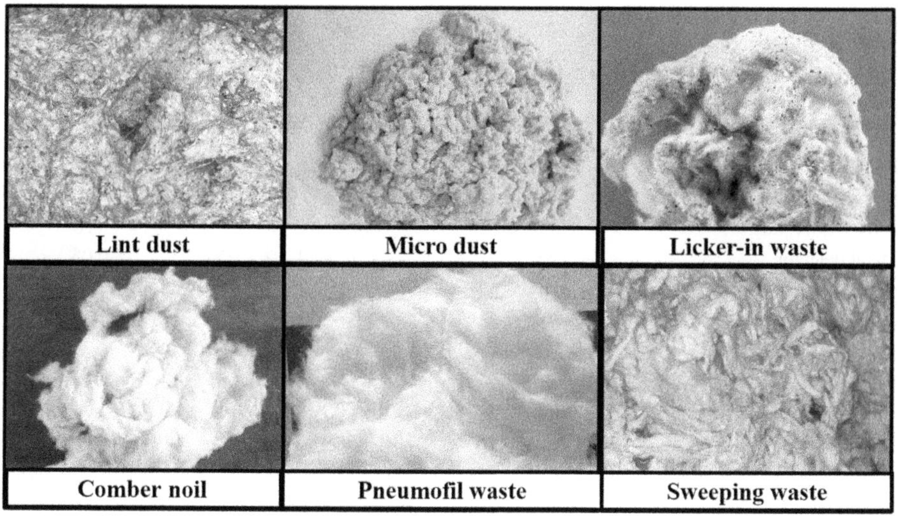

FIGURE 2.8 Different types of spinning wastes.

includes materials such as blow room droppings, licker-in droppings, flat strips, and noil (see Figure 2.8). It mainly consists of vegetable impurities like broken seed pieces, leaf fragments, short fibres, dust, and other similar contaminants. Microdust waste generated in cotton spinning mills is challenging to repurpose because of the extremely short fibre lengths and the difficulty of separating it from other particulate materials. In India, around 2.1 lakh tons of willow dust are produced, with open-end spinning units contributing approximately 3.5% to the microdust waste annually. In 2020, spinning mills generated an estimated 166.67 million kilograms of this microdust. At present, most of this waste is either sent to landfills or incinerated, highlighting the need for more sustainable waste management practices in the industry.

Higher spinning waste generation is observed in 100% cotton yarn production compared to cotton-polyester and polyester-viscose blends. This is due to a combination of fibre characteristics, processing requirements, equipment compatibility, and quality standards. Cotton fibres, while versatile and breathable, can be more challenging to spin efficiently due to their shorter staple length and higher propensity for lint formation. The shorter staple length of cotton fibres can lead to more fibre breakages during spinning, resulting in higher waste generation. Blending cotton with polyester fibres can improve the spinning process by adding strength and stability to the yarn. Polyester fibres are longer and more uniform than cotton fibres, resulting in fewer fibre breakages and less waste generation during spinning compared to 100% cotton yarn. Similarly, blending polyester with viscose fibres can also improve the spinning process, as viscose fibres are longer and smoother compared to cotton fibres. This can lead to lower waste generation during spinning compared to 100% cotton yarn production. Table 2.2 provides a detailed overview of the waste generated in the spinning segment for different fibre blends.

TABLE 2.2
Spinning Waste Percentage for Different Blends

Waste Type	100% Cotton	Cotton/Polyester	Polyester/Viscose
Blowroom			
Blowroom droppings	1.3–2	1–1.2	0.2–0.25
Kitty + microdust	0.6–1.2	0.8	0.1
Metal detector	0.05–0.1	0.1	0.05
Heavy particle	0.05–0.2	0.02–0.04	0
Contamination	0.5–1	0.3	0
Carding			
Flat waste + suction hood	2–4.5	2–2.15	1–1.5
Kitty + microdust	0.2–0.3	0.1–0.15	0
Licker-in	1.6–3	1.5–1.7	0.5–1
Sliver	0.2	0	0.5
Comber			
Noil	12–20	15–18 (cotton only)	0
Ring frame			
Pneumafil	1.5–2.5	0.5	1.6–1.7
Hard waste	0.1–0.2	0.1–0.15	0

2.3.3.2.1 Blowroom Waste

The blowroom process in textile manufacturing is essential for opening fibres into smaller tufts, cleaning them, and blending them to create a uniform mixture. This stage not only generates soft waste but also offers an opportunity for spinners to reintegrate this waste into the production process in appropriate proportions, thereby enhancing efficiency and sustainability. The quantity of waste collected in the blowroom is primarily influenced by the trash level present in the raw cotton fibre.

Several factors contribute to the quantity of waste produced in the blowroom, including the beater speed and settings, grid bars, pipeline suction pressure, and the speed and flow of the air current (40). Among these, the beater speed is the most significant factor. Higher beater speeds lead to more aggressive handling of the fibres, which can result in increased soft waste generation. This harsh action helps in liberating more waste material from the fibres but also leads to greater fibre damage and waste creation. Adjusting these parameters carefully is crucial to optimizing the balance between effective cleaning and minimizing fibre loss (41).

2.3.3.2.2 Carding Waste

The carding process involves fibre individualization from small tufts of fibres. The fibre sent out during individualization contributes to the generation of soft waste. At the carding stage, soft waste is produced from licker-in and cylinder droppings, as well as flats waste. The overall waste levels are also influenced by the cleaning efficiency of the blow room. Therefore, when evaluating waste generation, it is essential to consider the combined waste from both the blow room and carding sections.

Carding, in particular, produces a significant amount of waste, often referred to as "dirty waste," due to its high content of short fibres and contaminants. This stage is crucial for producing clean and parallel fibre slivers. Carding waste for cotton typically ranges from 5% to 6%, though it varies with the type of fibre; polyester waste is about 0.7–1%, and viscose is 1.1–1.6%. The waste from carding man-made fibres mainly comes from flat strips, with a metallic card's waste ranging from 1% to 2%, depending on the fibre type. The waste amount is influenced by the settings between the cylinder under casing and the cylinder, with optimal waste levels being 0.2% to 0.3%. If the length of flat waste exceeds 70% of the virgin fibre length, it indicates that good fibres are also being discarded, requiring adjustment in the flat setting. Smaller mote knives and wider feed plate-to-licker-in settings result in higher waste and lint loss. The carding process, which separates fibres from small tufts, generates soft waste. The waste from carding can contain 35–55% good fibre, depending on the quality of the initial fibre and the efficiency of the carding machine (42).

2.3.3.2.3 Comber Waste

The combing process enhances the quality of carded sliver by aligning fibres and eliminating neps. Key factors such as comber settings, top comb depth, and the extent of forward and backward feed determine the level of waste produced. This stage eliminates a large amount of short fibres as noil, some of which can be recycled back or used for manufacturing open-end yarns. Due to lower inter-fibre cohesion, combed slivers are more prone to breakage than carded slivers, resulting in higher sliver waste. The volume of waste generated during combing is largely influenced by the yarn count and specific requirements related to IPI (Imperfection Index), RKM (Resistance per Kilometre), and hairiness. Additionally, a higher proportion of comber waste does not necessarily indicate improved exclusion of short fibres and neps. Beyond a certain threshold, excessive waste may not contribute positively to yarn quality or performance. If the waste exceeds nominal levels, it leads to financial losses, while insufficient waste results in poor yarn quality and inconsistent count between bobbins.

Maintaining waste at a pre-determined level ensures uniformity in comber sliver wrapping, reducing the need for frequent checks. Combing, which enhances the quality of carded sliver by aligning fibres and removing neps, varies in intensity based on the desired quality class: semi-combed with less than 12% noil, normally combed with 12–18% noil, highly combed with 18–22% noil, and super-combed above 22%. Combing at noil levels below 12% is termed upgrading, improving cotton stock by eliminating short fibres, while above 22% is reserved for superfine yarns. In the past, forward feed was preferred for higher production rates with moderate quality demands, typically achieving noil percentages up to 14%. Modern high-performance combers now use forward feed across all staple lengths. This setup, combined with enhanced self-cleaning during fibre detachment and top combing, boosts retaining forces and allows noil extraction between 8% and 18%, improving overall waste control. Comber noil constitutes 16–20% of the comber waste and is a significant marketable product often used in the production of absorbent cotton, spunlace fabrics, and other non-woven products (43).

2.3.3.2.4 Draw Frame Waste

The draw frame process serves to attenuate sliver, eliminate hooks, blend fibres, and reduce yarn irregularities through doubling. During this stage, soft waste is produced, mainly in the form of broken sliver ends and fibre dust caused by high-speed operations. Additionally, unused sliver remaining at the end of a batch accounts for a significant portion of the soft waste. Common causes of waste at the draw frame include frequent end breakages, inappropriate spring pressure, and excessive or inadequate tension in the sliver at both the creel and delivery points. However, soft waste generation can be effectively reduced by ensuring proper pressure on drafting and calendar rollers.

In the spinning process, autoleveller waste refers to the waste material generated by the automatic levelling machines that are used to ensure uniform yarn thickness and quality. Autolevellers continuously monitor the thickness of the fibre sliver and make adjustments by either adding or removing material to achieve consistent yarn quality. The waste generated from this process primarily includes short fibres, uneven slivers, and small amounts of dust and impurities that are removed to maintain the desired yarn specifications. This waste is generally minimal in volume but can vary depending on the efficiency of the levelling system and the quality of the input material. The removed fibres and slivers, while typically short and less suitable for further spinning, may still have some value and can be recycled or used in lower-quality products, such as yarn for non-woven fabrics or as stuffing material.

2.3.3.2.5 Speed Frame Waste

The speed frame process attenuates the draw frame sliver into a suitable size for yarn formation by inserting a small amount of twist. Waste generated during this process is due to faults like cut roving, snapping, sloughing, and ratcheting. The root causes for these kinds of faults are bobbin jumping, broken tooth, irregular sliver feed, insufficient taper on bobbin, etc. Less than 0.5% of waste is generated from sliver and roving waste during speed frame operation.

2.3.3.2.6 Ring Frame Waste

The ring frame process includes drafting, twisting, and winding the yarn onto a bobbin. Waste levels tend to be higher primarily due to increased end breakages during spinning, which are influenced by the operational state and the material quality being processed. The quantity of waste generated is influenced by factors including the number of spindles per tender, the number of turns the tender completes on each spindle, and the quality of the roving fed. In mills, minimizing airborne waste is commonly a key focus of process control efforts.

2.3.3.2.7 Winding Waste

Usually, the waste occurring during winding stages is not fit for any further use and hence is classified as hard waste. Successful mill operation involves minimizing hard waste generation as much as possible. The amount of hard waste generated in the winding department should be kept less than 1%. Since this hard waste can fetch 1/4th of the amount of yarn price, the financial benefits of the mill are adversely affected when hard waste percentage is higher.

Autoconers are crucial in ensuring high-quality yarn production by efficiently managing cop rejection rates, ideally keeping them below 5%. To achieve this, it is essential to address several factors that contribute to bobbin rejection. Common issues include bobbin defects such as long tail ends, reshaped or overfilled bobbins, handling stains, neppy yarns, improper winding tension, broken ends, and various types of damage (see Figure 2.9). Issues may also occur due to bobbin feeding problems in the magazine, failures in top bunch transfer, faults in the winding unit, and yarn quality defects. Furthermore, problems such as double gaiting or excessive piecing from the ring frame, inadequate suction in the gripper arm, and the use of damaged empty tubes should be prevented. To minimize waste, rejected bobbins should be rewound on conventional cone winding machines at slow speeds after defect removal, rather than cutting them. Immediate attention to red light alerts, proper yarn tensioning, and minimizing splicing lengths are crucial for maintaining efficiency. Regular maintenance of records and periodic reconciliation of hard waste figures with warehouse records are also necessary. Modern Autoconers are equipped with settings to reject bobbins with yarn counts exceeding certain limits, which can make processing yarns with even slight count variations challenging.

2.3.3.3 Weaving Industry

The weaving industry involves the process wherein the yarn produced from the fibre is converted into fabric using two sets of yarn, namely warp yarn and weft yarn, respectively. During this yarn-to-fabric formation, certain amounts of waste

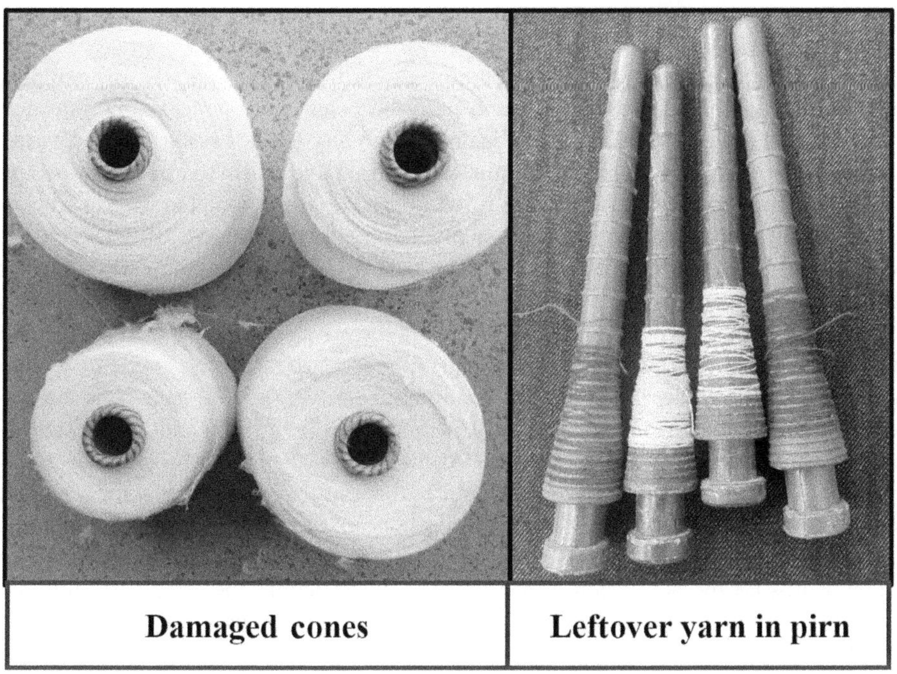

FIGURE 2.9 Package wastes generated from winding process.

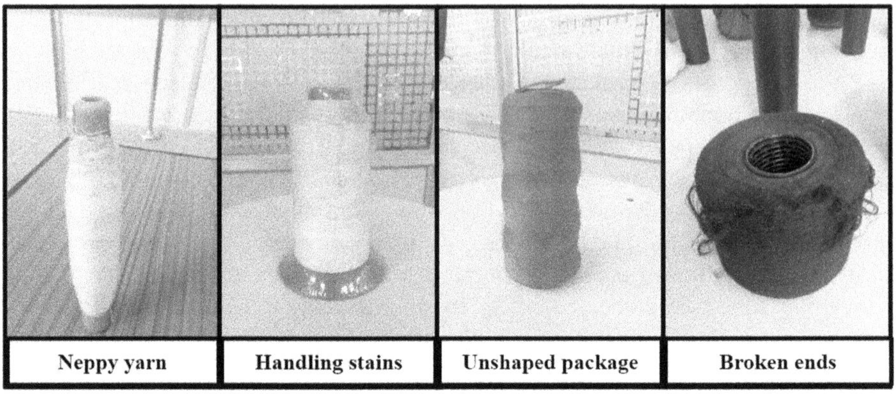

FIGURE 2.10 Wastes generated from weaving industry.

are produced during different sections of the weaving processes (see Figure 2.10). Residual yarn remains on cones after the warping process, as it is not feasible to fully deplete every cone in the warping creel. A small portion of yarn typically stays behind on each cone. The number of short stops in weaving is a key quality indicator for the yarn's weaving ability. For 100% combed cotton yarn, an acceptable rate is 1.5 short stops per 100,000 metres of yarn due to weft breaks.

2.3.3.3.1 Sizing Waste

Sizing waste is a type of waste generated in weaving factories. In the weaver's beam section, initiating a new warp set requires discarding some threads to guarantee that only properly sized yarn is wound onto the beam. Following sizing, knotting is performed to join the ends of two beams, which also creates some leftover waste. Another form of hard waste is residual waste from the warp beam, which occurs when a minimal portion of warp yarn is left unused on the completed weaver's beam and cannot be fully recovered. Additionally, auxiliary selvage—an artificial selvage used to hold the weft yarn during the loom beat-up phase—results in waste during the weaving process.

Waste is produced at two main areas during the sizing process: the creel side and the headstock side. When changing or adjusting the warp set, yarn is pulled from the creel until the sheet becomes aligned, resulting in waste. Overall, waste from the sizing process typically ranges between 0.5% and 1%, though this amount is significantly affected by the yarn count used. At the start of the sizing process, an initial portion of warp yarn is discarded to guarantee that only properly sized yarn is wound onto the weaver's beam. Additionally, a considerable length of warp sheet remains unused at the end of each sizing cycle and is treated as waste, since it is needed to tie in the warps for the next batch. When the new sizing begins, the sized yarn from the old batch is removed and discarded. In addition, a significant part of the warp threads is left as residual waste on the beams.

2.3.3.3.2 Warping Waste

During warping, it is generally impractical to completely use up a yarn cone or cheese package, resulting in leftover yarn residues. This accounts for hard waste being generated. As the cone package nears depletion, its decreasing diameter leads to an increase in the number of unwinding coils. This change makes it difficult to maintain a constant warping speed, often resulting in warp yarn breakage. This breakage results in hard waste generation. Moreover, since the yarn length on each cone varies, the cones on the warping creel do not run out simultaneously. This leads to leftover yarn on individual cones, contributing to material waste (44).

In warping, the primary goal is to minimize missing ends, with the target being no more than seven thread breakages per 10 million metres of yarn. Ideally, the number of end breaks per million metres should be 5 for carded yarn and 3 for combed yarn. Achieving this requires maintaining high-quality yarn packages and adhering to specific process parameters, including precise creel alignment, reliable stop motions on the creel and warping drum, minimal wobbling of warping beams, and uniform selvage with good flanges. Yarns should be warped on spindle-driven machines to prevent thermal damage from abrasion. Despite these measures, some hard waste is unavoidable due to residues left in cone or cheese packages, increased unwinding coils as the cone diameter decreases, and variable yarn lengths leading to residual yarn on individual cones.

2.3.3.3.3 Knotting Waste

Knotting is performed when the total count of warp and the weave of the old and new beams are identical. This method ties knots on a "one-to-one" basis between the existing warp ends on the loom and those on the new weaver's beam, streamlining the process by eliminating the need for beam gaiting and drawing-in. Trimming the short sections of warp sheet from both the new beam and the existing ends generates knotting waste. Typically, acceptable hard waste from knotting ranges between 0.3% and 0.5%, which is manageable, but losses increase if this percentage rises. Therefore, controlling knotting waste is important when assessing total waste in an industry. The amount of knotting waste can vary based on worker practices and the number of knots tied.

2.3.3.3.4 Gaiting Waste

Gaiting waste occurs when installing a new warper beam that differs in the number of ends or fabric design from the previous beam. After gaiting, the drawing-in and denting processes follow, where warp yarns are threaded through the appropriate parts. Gaiting or tying is also required when setting up a new loom. Typically, the waste generated during gaiting ranges between 0.3% and 0.5%.

2.3.3.3.5 Auxiliary Selvedge Waste

Auxiliary selvedges are a set of dummy selvedges used to support the pick during the beat-up process (see Figure 2.11). Typically consisting of 5 to 9 warp yarns, auxiliary selvedges are arranged at a specific distance from the main fabric's outermost warp

FIGURE 2.11 Auxiliary selvedge wastes from weaving process.

threads. This selvedge is later removed by cutting with scissors and mainly functions to support the pick in place during the reed movement. Once beat-up is complete, the auxiliary selvedge is separated from the body fabric and gathered in a designated waste bin. Typically, selvedge waste accounts for about 1.5% to 2.0% of the total material and is categorized as hard waste.

2.3.3.4 Knitting Industry

The primary objective of the knitting and weaving process is quite similar, wherein the conversion of yarn from fibre to fabric is the main goal. The main difference lies in the mechanism by which the fabric is formed. Knitted fabrics are created by the inter-looping of yarns. Knitted materials contribute a large portion to the total textiles produced globally and hold a significant market share economically. The knitting process includes wastage such as trial run quantity, leftover yarn in cones and yarn in the machine tube, hole formation, knitting defects, roll making, etc. (see Figure 2.12).

2.3.3.4.1 Trial Run Waste

To process a new order in bulk quantities, initially, the company conducts trial runs of the fabric with the specifications mentioned in the tech pack developed by the buyer.

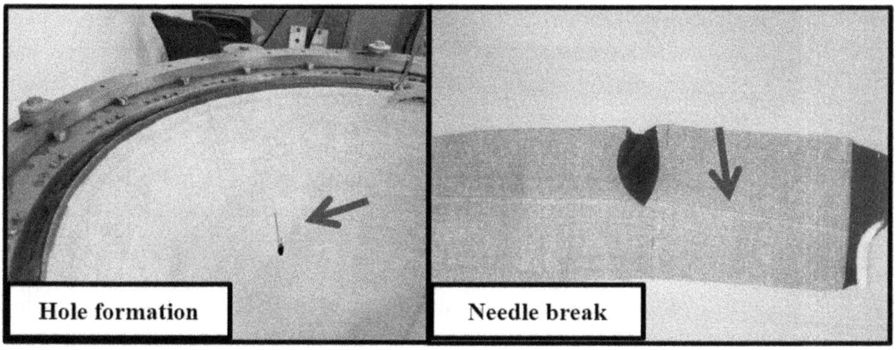

FIGURE 2.12 Wastes generated from knitting industry.

In some cases, the sample might not match the exact requirements specified by the buyer, and this sample is considered waste. It falls under the category of soft waste. The waste generated from the trial run can be reused in the same industry for other purposes. The total wastage occurring due to sample trials is typically in the range of 0.5–1%.

2.3.3.4.2 Winding Waste

In the preparation of knitted fabric, the process shares similarities with weaving, where yarns are fed into the knitting machine through cone packages. Yarn that remains in the cone packages and is not fully utilized generates hard waste. This waste results from residual yarn left over as the cones are emptied. Fly, which consists of small fibres or dust, can accumulate and lead to hard waste. This problem typically arises from improper settings or defective yarn guides, which transport yarn from the cone package to the knitting machine. The generation of fly waste contributes about 1% to the total wastage during the knitting process. Breaks in the yarn, often caused by incorrect tension settings or machine faults such as issues with the package, lead to additional waste. Effective management of yarn tension and regular maintenance of machinery are essential to minimizing these breaks and thus reducing overall waste.

2.3.3.4.3 Knitting Waste

The knitted fabric developed may encounter various faults such as spiralling, shrinkage, pinholes, and neps. These faults occur due to variations in yarn properties used, poor machine maintenance, improper machine settings, tension variation, etc. Improper cone shape can affect the yield of the cone, leading to a large amount of wasted yarn. Therefore, maintaining proper cone shape is mandatory. Around 2–3% of yarn from the cone gets wasted due to improper shaping of the cone. To minimize this loss, the shape of the cone must be checked and analyzed before feeding it into the knitting creels. Cone package faults in the knitting industry, such as yarn contamination (1–5%), hard or soft cones (5–10%), slubs and neps (2–8%), cone collapse (1–3%), yarn breaks (2–7%), and tangling (2–5%), can significantly impact production efficiency and fabric quality. These faults typically affect 1–10% of cones, depending on factors like the quality of raw materials, equipment maintenance, and winding process precision. Yarn faults cause the majority of yarn wastage on the knitting floor. Faulty yarns tend to break easily, leading to additional waste during knotting. Approximately 4–5% of yarn is lost due to these defects. To reduce wastage from yarn faults, it is important to use fault-free yarn. Combed yarn is preferred over carded yarn since it contains fewer short fibres, resulting in better quality and less waste.

2.3.3.5 Textile Processing Industry

The manufacturing of textile products involves huge amounts of chemical processing to achieve finished products. Some of the main chemical processes involved are desizing, scouring, bleaching, mercerization, dyeing, printing, and finishing. Each process entails significant usage of water and chemicals, leading to the release of effluents that cause significant stress on the well-being of the ecosystem. Textile industry wastewater contains numerous pollutants, including acids, heavy metals, bases, toxic organic and inorganic dissolved substances, as well as dyes. Colours, primarily stemming from various dyes, are considered the most undesirable substance (45,46).

Dyes are discharged into the environment by textile industries and dye production facilities (47). Dyeing and printing are the most water-intensive stages in textile processing. The wastewater produced from these processes is generally high in organic matter, dyes, and heavy metals (48). The wastewater also carries remnants of different dyes and chemicals that are harmful to aquatic organisms and microbes, highlighting the need for thorough treatment before disposal. Conventional textile finishing processes can use approximately 100 L of water to process just 1 kg of textile material (49). A small industry producing 8,000 kg per day requires approximately 1.6 million litres of water. About 16% of water is used for dyeing, with half of that amount allocated for printing. During conventional dyeing and finishing processes, 200,000 L of water become contaminated per 1,000 kg of fabric.

In general, the quantity of waste generated in textile wet processing can be reduced at different stages such as fabric preparatory stages, fabric dyeing stages, and fabric finishing stages. In the fabric preparation stage, waste is minimized by the recovery of heat and chemicals, and the use of alternative eco-friendly materials as reagents instead of conventional ones. In dyeing operations, reuse of the same dye bath can reduce the amount of waste liberated, and the use of green and natural dyes is to be encouraged.

Textile processing generates a variety of wastes, each with distinct characteristics and environmental impacts. During desizing, substances such as starch, glucose, polyvinyl alcohol (PVA), and resins that were applied to the fabric during the sizing process are removed, producing a waste stream with a very small volume but high biological oxygen demand (BOD), particularly from PVA, contributing 30–50% of the total BOD. Scouring introduces caustic soda, waxes, and sodium silicate, yielding a dark-coloured, strongly alkaline waste with significant BOD, accounting for 30% of the total. Bleaching, which uses hypochlorite, chlorine, and caustic soda, generates small volumes of strongly alkaline waste with a low BOD (around 5%). Mercerization, involving caustic soda, results in similarly small, alkaline waste streams with minimal BOD. Dyeing, a more voluminous process, produces highly coloured effluents with fairly high BOD due to the presence of dyes and mordant. Printing adds to this with a small, oily waste stream that also exhibits a relatively high BOD. Finally, finishing processes leave behind traces of starch, tallow, and other chemicals, resulting in a less alkaline, low BOD waste. Overall, the combination of these processes contributes to the textile industry's environmental footprint, particularly in terms of chemical and organic waste load.

2.3.3.6 Garment Industry

The garment industry is a labour-intensive sector that requires a large workforce to produce the required apparel from fabrics obtained from weaving or knitting industries. The apparel sector, valued at $1.8 trillion, manufactures 150 billion pieces of clothing annually, averaging almost 19 garments per person worldwide (16). Garment production involves numerous processes, and efficient management is essential for making a company profitable.

The fashion industry produces an astonishing 13 MT of waste each year. Waste generated in the garment sector can be classified into three main types: processing waste, packing waste, and accessories waste. During garment manufacturing, fabric waste accounts for approximately 15% to 30% of the total raw material input. When not repurposed or recycled, these scraps usually end up in landfills. The initial waste in apparel manufacturing originates from marker planning, where patterns are created for cutting and sewing the garments (see Figure 2.13). For example, when producing a basic shirt using traditional pattern-making methods, only about 85% of

FIGURE 2.13 Cutting wastes generated from garment industry.

the fabric is utilized, leaving approximately 15% as waste in the cutting area (50). All this fabric waste contributes to a "significant ecological footprint."

A study on fabric waste in T-shirt production revealed that the cutting process generates the most waste, averaging 13.57%, followed by 6.91% in the control panel, 4.31% during sewing, and 1.72% in the finishing stage, resulting in a total waste of 26.51% of the original fabric. Around 5.7 lakh tons of waste is reportedly generated annually by Bangladesh's garment industry. Of this, nearly 250,000 tons—close to half—is pure cotton waste that is fully recyclable (51). Research indicates that 2,238 sustainable products can be produced using 218.6 kg of surplus fabric, 212.13 kg of reusable cutting waste, and 210 rejected cut panels suitable for reuse (52).

2.3.4 Waste Categorization Based on Defects

A defect is an imperfection that reduces the quality or performance of the material. Based on defects, textile waste is classified into three categories: yarn defects, fabric defects, and garment defects.

2.3.4.1 Yarn Defects

Yarn defects refer to imperfections that occur during the spinning or yarn-making process. These defects can include irregularities in the thickness or diameter of the yarn, presence of foreign matter within the yarn, uneven twisting, knots, slubs (thickened areas), or thin places. Yarn defects can compromise the strength, durability, and appearance of the fabric produced from them. Additionally, yarn defects can propagate and become more pronounced as the yarn is further processed into fabric and garments (53).

The most common yarn defects include thick places, thin places, uneven surfaces, and slubs, which together make up about 72% of all yarn defects. These imperfections not only compromise yarn quality but also negatively impact subsequent manufacturing processes and the final product (54). Thick places are sections where the yarn diameter is larger than adjacent areas and typically measure around 6 mm (¼ inch) in length, often caused by improper fibre drafting during yarn formation. Conversely, thin places are sections where the diameter is significantly smaller—by at least 25%—compared to the average yarn diameter, usually resulting from inconsistencies in the sliver or yarn count (55).

2.3.4.2 Fabric Defects

Fabric defects occur during the weaving or knitting process, where yarns are interlaced or looped to form fabric. These defects can manifest as holes, tears, stains, uneven weaving or knitting, misalignment of yarns, or pattern irregularities (see Figure 2.14). Fabric defects not only affect the aesthetics of the fabric but also its structural integrity and performance characteristics such as tensile strength, tear resistance, and drape. Defective fabric may need to be discarded or repaired, leading to increased material waste and production costs.

Yarn defects can significantly impact the quality and usability of textile products. Common defects include drop stitches caused by high yarn tension or obstructions,

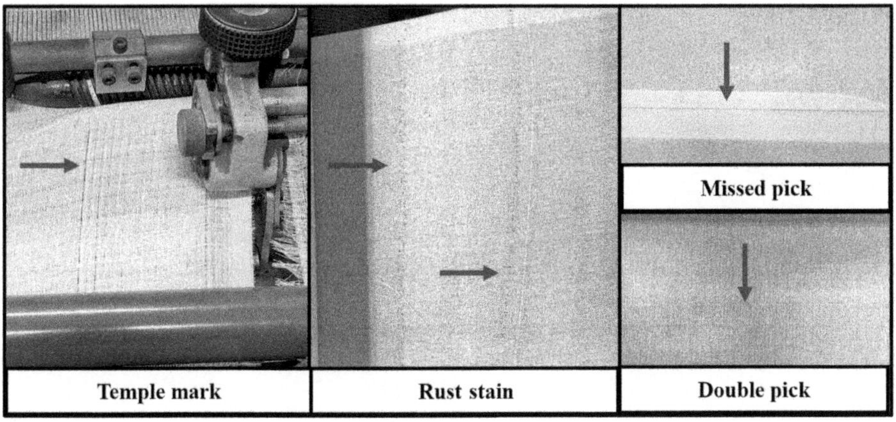

FIGURE 2.14 Different types of fabric wastes generated due to defects.

barriness from yarn count variations or package hardness inconsistencies, and streakiness due to yarn slippage or faulty winding. Snarls arise from high yarn twist or improper tension, while contamination occurs from dead fibres or foreign matter. Spirality is a result of high twist per metre discrepancies, and bowing is due to uneven tension distribution during finishing. Oil lines are caused by excessive oiling and fibre accumulation in machine parts, whereas surface hairiness and piling stem from abrasion and rough machine surfaces. Proper maintenance, tension control, and cleaning are essential remedies for these defects.

2.3.4.3 Garment Defects

Garment defects occur during the garment manufacturing process and can result from issues such as poor cutting, sewing, or finishing techniques. Common garment defects include stitching errors, misaligned seams, loose threads, puckering, fabric distortion, sizing inconsistencies, and faulty closure (see Figure 2.15). Garment defects not only affect the appearance and fit of the final product but also its functionality and durability. Defective garments may need to be reworked or scrapped, leading to wasted materials, labour, and resources.

Adult outerwear garments typically produce waste ranging from 10% to 20%, with trousers or pants generating around 10%, while blouses and jackets tend to have higher waste percentages. This is mainly because the irregular shapes of pattern pieces prevent perfect nesting, making full fabric utilization impossible (56). Overall, approximately 15% to 35% of raw materials used in garment production end up as factory waste, depending on the type of product. On average, about 15% waste is generated during the design phase through pattern cutting and marker planning.

Another major factor contributing to garment waste is waste generated due to sewing errors, which are human-made errors and can be minimized only by proper training and the use of adequately skilled labour. Since these wastes mainly consist of small fabric pieces, they are mostly unusable and fall under the category of hard waste. Waste also arises due to improper size matching, shade variation, and other defects in the garment parameters assessed during quality control assessments.

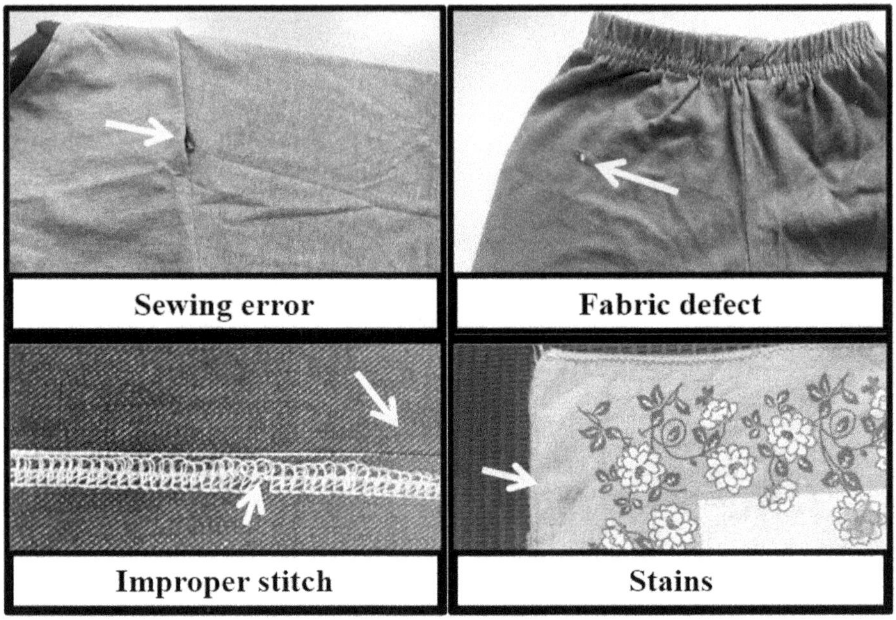

FIGURE 2.15 Different types of garment wastes generated due to defects.

In some cases, fabric rolls may be available in surplus even after completing the required job order; this may be considered soft waste since it can be resold or reused in other useful applications.

2.3.5 Waste Categorization Based on Toxicity

Based on the safety aspect of the liberated wastes, they are divided into two categories: hazardous and non-hazardous waste.

2.3.5.1 Hazardous Wastes

Hazardous waste refers to materials that pose serious threats to human health or the environment if they are not managed and disposed of correctly. The Environmental Protection Agency classifies a substance as hazardous waste if it appears on specific hazardous waste lists or exhibits certain hazardous characteristics. Regulatory oversight of hazardous waste comes under the Resource Conservation and Recovery Act (RCRA), a law of the United States. Accordingly, different countries implement their own regulatory frameworks, as presented in Table 2.3.

Textile manufacturing processes like bleaching and mercerizing produce hazardous waste including hydrogen peroxide, sodium silicate, alkali, and sodium hydroxide. Additionally, equipment maintenance yields hazardous substances like methyl ethyl ketone, tetrachloroethylene, trichloroethylene, toluene, isopropyl alcohol, and mineral spirits (57). Untreated textile effluents from various textile processes such as dyeing and printing are classified as hazardous waste, posing an immediate threat

TABLE 2.3
Textile Hazardous Waste Regulatory Acts by Country

Country	Regulation/Act
United States	Resource Conservation and Recovery Act (RCRA), 1976
European Union	Waste Framework Directive (2008/98/EC) and REACH Regulation (EC No. 1907/2006)
India	Hazardous and Other Wastes Rules, 2016 under the Environment Protection Act
China	Law on the Prevention and Control of Environmental Pollution by Solid Waste
Bangladesh	Hazardous Waste and Ship Waste Management Rules, 2011 (provisional) and Environment Conservation Rules, 1997
Vietnam	Law on Environmental Protection, 2020
Turkey	Regulation on Control of Hazardous Wastes (based on EU Waste Directive)
Indonesia	Government Regulation No. 101/2014 on Hazardous and Toxic Waste Management
Sri Lanka	National Environmental Act, No. 47 of 1980

to land and marine ecosystems. Textile industry wastewater is highly alkaline and has large amounts of BOD, chemical oxygen demand, total dissolved solids, and alkalinity. If not properly treated before being released, it can cause environmental problems.

Effluents from textile processes contain a variety of harmful substances including soap, salts, dyes, acids, chromium compounds, and heavy metals such as copper, mercury, and cobalt, along with various auxiliary chemicals, making the wastewater highly toxic. Other dangerous chemicals found in the water include formaldehyde, hydrocarbons, and non-biodegradable dyeing agents. Around 40% of the colourants used globally have organically bound chlorine, which is known to cause cancer. These chemicals can either evaporate into the air or be absorbed through the skin, leading to allergies and serious health hazards (56).

Healthcare products such as surgical masks, gowns, caps, and sanitary napkins contaminated with blood are considered hazardous waste. The destination of these kinds of waste is incineration. Microplastics, frequently stemming from the washing of synthetic fabrics, heavily contaminate the environment, with more than 14 MT per year settling on the ocean floor. Synthetic textiles are responsible for 16–35% of the world's microplastic pollution, releasing between 200,000 and 500,000 tons each year, especially from fast fashion products. Mitigating these emissions is achievable through sustainable design, enhanced production techniques, and improved disposal practices (58).

2.3.5.2 Non-Hazardous Textile Waste

Non-hazardous textile waste does not cause immediate harm to the environment upon exposure to the external surroundings. It includes commonly used waste fibres, waste fabric pieces, and yarn fragments, which may or may not be reused in further applications. This waste accounts for the majority of textile waste, with hazardous waste constituting only 10–15% of total industrial waste production. The generation of hazardous textile waste is growing annually at a rate of 2–5% (1). Uncoloured textile waste originates from various stages of the textile production process, including knitting and finishing. Coloured textile waste primarily consists of scrap products from dyeing units.

2.3.6 Waste Categorization Based on Biodegradability

Based on the physical aspects, the wastes can be categorized into two categories—compostable wastes and recyclable wastes.

2.3.6.1 Compostable Waste

Textile waste procured from natural fibres such as cotton, wool, silk, etc. is considered compostable material where composting can be touted as a viable alternative to conventional landfilling of the waste. However, this process is applicable only to materials that are fully biodegradable and not blended with any synthetic materials. Growing awareness of the environmental harm caused by non-biodegradable synthetics has spurred interest in degradable textiles. Products like nappy pads, wipes, and car interiors are now designed for biodegradability. Non-wovens and disposables are increasingly popular. Industry research focuses on 100% natural materials that decompose in landfills, including natural and regenerated fibres. PLA biodegradable polymers, derived from corn, offer promise, especially for medicinal textiles due to their antimicrobial properties. Studies explore converting post-industrial waste into composts for use as bio-manure, with effective microorganisms enhancing soil, plant, and water health (59).

The impact of dyes and finishes on compostability is significant, as many commonly used chemicals in textiles can hinder the natural degradation of fibres and introduce toxic substances into compost. Synthetic dyes (e.g., azo dyes, disperse dyes) can contain heavy metals and aromatic amines that resist biodegradation and may be toxic to soil microorganisms. Finishes such as water repellents, flame retardants, antimicrobials, and wrinkle-free coatings often involve fluorocarbons, formaldehyde, or silicone-based compounds, which reduce the biodegradability of fabrics and contaminate compost. The presence of these chemicals can slow down composting rates, affect microbial activity, and compromise the quality of the final compost product. Natural dyes and biodegradable finishes (e.g., chitosan-based or plant-derived resins) have shown better compatibility with composting processes.

2.3.6.2 Recyclable Waste

Recyclable textile waste includes both biodegradable materials like cotton and wool, as well as non-biodegradable materials such as synthetic fibres and blends, provided they can be processed for reuse or material recovery. Waste from synthetic materials is highly unlikely to decompose completely; therefore, it is taken for recycling and reformed into another useful product, not necessarily a textile product itself. Approximately 80% of recyclable waste lacks proper segregation, requiring manual sorting. Upon reaching collectors and traders, 40% is discarded due to contamination or mismatch with buyers. Despite this, about 70% of post-industrial textile waste likely enters new life cycles through industrial symbiosis. However, there is no tracking system for its destination. The USA alone generates 11.9 MT of PCTW yearly, possibly totalling 84 MT globally. Around 30% may be collected, with half potentially resold. Thus, a maximum of 12 MT of PCTW could feasibly fuel the recycling industry worldwide (60).

2.4 SUMMARY

The textile industry encompasses a complex array of processes, each contributing to the generation of waste resources that pose significant environmental and economic challenges. Through an overview of textile processes and waste resources across various categories, including spinning, weaving, knitting, processing, and garments, it becomes evident that waste management and sustainability are crucial considerations for the industry's future.

The origin of textile waste is diverse, ranging from pre-consumer waste generated during manufacturing faults to post-consumer waste resulting from discarded clothing and industrial waste from non-traditional textile applications. Each category presents unique challenges for waste management and disposal, highlighting the need for comprehensive strategies to address waste throughout the supply chain.

Moreover, the safety aspect of textile waste is a significant concern, with hazardous waste containing chemicals, dyes, and heavy metals posing risks to human health and the environment. Non-hazardous waste, while less immediately threatening, still requires proper management to prevent environmental pollution and resource depletion. Furthermore, the physical properties of textile waste play a crucial role in determining its disposal and recycling potential. Compostable waste derived from natural fibres offers opportunities for eco-friendly disposal through composting, while non-biodegradable recyclable waste, such as synthetic fibres, can be repurposed into new products through recycling initiatives.

In conclusion, the textile industry must adopt a multifaceted approach to address waste resources effectively. This includes implementing quality control measures to minimize defects, promoting circular economy principles to reduce waste generation, and investing in sustainable technologies and practices across the supply chain. Focusing on reducing waste, recycling, and caring for the environment allows the textile industry to lessen its impact and move towards a more sustainable future.

REFERENCES

1. Agrawal R, Sharan M. Municipal textile waste and its management. *Research Journal of Family, Community and Consumer Science.* 2015 Mar;3(1):4–9.
2. La Rhea Pepper, Liesl Truscott. Preferred fiber & materials, market report 2020. Textile Exchange [Internet]. [cited 2021 Jan 10]; Available from: https://store.textileexchange.org/product/2020-preferred-fiber-materials-report/
3. Juanga-Labayen JP, Labayen IV, Yuan Q. A review on textile recycling practices and challenges. *Textiles.* 2022 Mar 16;2(1):174–88.
4. Sumanta Bhattacharya. Impact of the new textile policy and textile waste management system in India and a move towards sustainable management. *International Journal for Research in Applied Science and Engineering Technology.* 2021 Nov 30;9(11):1016–21.
5. Ringler C, Zhu T. Water resources and food security. *Agronomy Journal.* 2015 Jul;107(4):1533–8.

6. Leal Filho W, Ellams D, Han S, Tyler D, Boiten VJ, Paço A, et al. A review of the socio-economic advantages of textile recycling. *Journal of Cleaner Production.* 2019 May;218:10–20.
7. ECOSIGN. Textile recycling as a contribution to circular economy and production waste enhancement. 2017. Available from: http://www.ecosign-project.eu/news/textile-recycling-as-a-contribution-to-circular-economy-and-production-waste-enhancement/
8. Kasavan S, Yusoff S, Guan NC, Zaman NSK, Fakri MFR. Global trends of textile waste research from 2005 to 2020 using bibliometric analysis. *Environmental Science and Pollution Research.* 2021 Sep;28(33):44780–94.
9. IMARC Group. India textile recycling market: Industry trends, share, size, growth, opportunity and forecast 2023–2028. IMARC, 2023. Available from: https://www.imarcgroup.com/india-textile-recycling-market
10. Statista [Internet]. Distribution of the global population 2023, by continent. 2023. Available from: https://www.statista.com/statistics/237584/distribution-of-the-world-population-by-continent/
11. Statista [Internet]. Production volume of chemical and textile fibers worldwide from 1975 to 2022. 2022. Available from: https://www.statista.com/statistics/263154/worldwide-production-volume-of-textile-fibers-since-1975/
12. Morlet A, Opsomer R, Gillet C, Herrmann S, Balmond L, Fuchs L. A new textiles economy: redesigning fashion's future [Internet]. Ellen MacArthur Foundation; Available from: http://www.ellenmacarthurfoundation.org/publications
13. Shakya A. Integrated waste minimization techniques in apparel design: a sustainable perspective [Internet]. Dayalbagh Educational Institute; 2022. Available from: http://hdl.handle.net/10603/489532
14. Vyas H. Creating the vogue through sustainability: Waste management in the garment industry [Internet]. OCAD University, 2022. Available from: https://openresearch.ocadu.ca/id/eprint/3797/1/Vyas_Hetvi_2022_MDes_SFI_MRP.pdf
15. Azad AK, Haq UN, Khairul Akter MMd, Uddin MA. Recycling practices of pre-consumer waste generated from textile industry. In: Muthu SS, editor. *Sustainable Manufacturing Practices in the Textiles and Fashion Sector [Internet].* Cham: Springer Nature Switzerland; 2024 [cited 2024 May 16]. p. 301–24. (Sustainable Textiles: Production, Processing, Manufacturing & Chemistry.) Available from: https://link.springer.com/10.1007/978-3-031-51362-6_12
16. De Silva R. Waste isn't waste until we waste it: Repair as a best-practice in reducing pre-consumer industrial fashion waste [Internet]. European Union; 2023. Available from: https://circulareconomy.europa.eu/platform/sites/default/files/2023-10/WearnRepair_casestudy
17. Parisi ML, Fatarella E, Spinelli D, Pogni R, Basosi R. Environmental impact assessment of an eco-efficient production for coloured textiles. *Journal of Cleaner Production.* 2015 Dec;108:514–24.
18. Echeverria CA, Handoko W, Pahlevani F, Sahajwalla V. Cascading use of textile waste for the advancement of fibre reinforced composites for building applications. *Journal of Cleaner Production.* 2019 Jan;208:1524–36.
19. Shirvanimoghaddam K, Motamed B, Ramakrishna S, Naebe M. Death by waste: fashion and textile circular economy case. *Science of the Total Environment.* 2020 May;718:137317.
20. Ellen MacArthur Foundation. Fashion and the circular economy [Internet]. [cited 2022 Feb 20]. Available from: https://ellenmacarthurfoundation.org/topics/fashion/overview2018

21. Sinha P, Dissanayke DGK, Abeysooriya RP, Bulathgama BHN. Addressing post-consumer textile waste in developing economies. *The Journal of The Textile Institute*. 2022 Sep 2;113(9):1887–907.
22. Akber Abbasi S, Khalil AB, Arslan M. Extensive use of face masks during COVID-19 pandemic: (Micro-)plastic pollution and potential health concerns in the Arabian Peninsula. *Saudi Journal of Biological Sciences.* 2020 Dec;27(12):3181–6.
23. Tripathi A, Tyagi VK, Vivekanand V, Bose P, Suthar S. Challenges, opportunities and progress in solid waste management during COVID-19 pandemic. *Case Studies in Chemical and Environmental Engineering.* 2020 Sep;2:100060.
24. Torres FG, De-la-Torre GE. Face mask waste generation and management during the COVID-19 pandemic: An overview and the Peruvian case. *Science of the Total Environment.* 2021 Sep;786:147628.
25. Xiang Y, Song Q, Gu W. Decontamination of surgical face masks and N95 respirators by dry heat pasteurization for one hour at 70°C. *American Journal of Infection Control.* 2020 Aug;48(8):880–2.
26. Kumar T, Mishra S, Verma RK. Fabrication and tensile behavior of post-consumer carpet waste structural composite. *Materials Today: Proceedings.* 2020;26:2216–20.
27. Sotayo A, Green S, Turvey G. Carpet recycling: A review of recycled carpets for structural composites. *Environmental Technology & Innovation.* 2015 Apr;3:97–107.
28. Cunningham PR, Miller SA. A material flow analysis of carpet in the United States: Where should the carpet go? *Journal of Cleaner Production.* 2022 Sep;368:133243.
29. Mazharul Islam Kiron. List of waste in spinning mill (blowroom, carding, drawframe, combing, simplex and ring frame) [Internet]. 2012. Available from: https://textilelearner.net/list-of-waste-in-spinning-mill/
30. Mishra PK, Izrayeel AMD, Mahur BK, Ahuja A, Rastogi VK. A comprehensive review on textile waste valorization techniques and their applications. *Environmental Science and Pollution Research.* 2022 Sep;29(44):65962–77.
31. Plácido J, Imam T, Capareda S. Evaluation of ligninolytic enzymes, ultrasonication and liquid hot water as pretreatments for bioethanol production from cotton gin trash. *Bioresource Technology.* 2013 Jul;139:203–8.
32. Cheng HN, Dowd MK, Finkenstadt VL, Selling GW, Evangelista RL, Biswas A. Use of cotton gin trash and compatibilizers in polyethylene composites. In: Cheng HN, Gross RA, Smith PB, editors. *ACS Symposium Series [Internet].* Washington, DC: American Chemical Society; 2013 [cited 2023 Dec 1]. p. 423–31. Available from: https://pubs.acs.org/doi/abs/10.1021/bk-2013-1144.ch028
33. Hamawand I, Sandell G, Pittaway P, Chakrabarty S, Yusaf T, Chen G, et al. Bioenergy from cotton industry wastes: a review and potential. *Renewable and Sustainable Energy Reviews.* 2016 Dec;66:435–48.
34. Haque ANMA, Remadevi R, Naebe M. A review on cotton gin trash: Sustainable commodity for material fabrication. *Journal of Cleaner Production.* 2021 Jan;281:125300.
35. Sutivisedsak N, Cheng HN, Dowd MK, Selling GW, Biswas A. Evaluation of cotton byproducts as fillers for poly(lactic acid) and low density polyethylene. *Industrial Crops and Products* 2012 Mar;36(1):127–34.
36. Singh S, Kumar A. Soft waste management in spinning industry. In: Agnihotri AK, Reddy KR, Bansal A, editors. *Recycled Waste Materials [Internet].* Singapore: Springer Singapore; 2019 [cited 2024 May 16]. p. 213–7. (Lecture Notes in Civil Engineering; Vol. 32.) Available from: http://link.springer.com/10.1007/978-981-13-7017-5_23
37. Bangladesh Textile Mills Association (BTMA). Basic data of primary textile sector [Internet]. 2020. Available from: https://www.btmadhaka.com/wp-content/uploads/2021/04/Basic-Data-of-BTMA-2020

38. Alagirusamy R. Process control in blowroom and carding operations. In: Majumdar A, Alagirusamy ADR, and Kothari VK, editors. *Process Control in Textile Manufacturing [Internet]*. Cambridge, UK: Elsevier; 2013 [cited 2024 May 16]. p. 132–57. Available from: https://linkinghub.elsevier.com/retrieve/pii/B9780857090270500060
39. Nityanand AR. Waste investigation and control for a spinning mill. *Indian Textile Journal.* 1984 Sep;9(3):79–86.
40. Thilagavathi G, Karthik T. *Process Control and Yarn Quality in Spinning*. New Delhi: Woodhead Publishing India Pvt Ltd; 2015. 410 p. (Woodhead Publishing India in Textiles.)
41. Lawrence CA, Textile Institute, editors. *Advances in Yarn Spinning Technology*. Oxford: Woodhead Publishing; 2010. 446 p. (Woodhead Publishing Series in Textiles.)
42. Bedez Ute T, Celik P, Bunyamin Uzumcu M. Utilization of cotton spinning mill wastes in yarn production. In: Körlü A, editor. *Textile Industry and Environment* [Internet]. Intech Open; 2019 [cited 2024 May 16]. Available from: https://www.intechopen.com/books/textile-industry-and-environment/utilization-of-cotton-spinning-mill-wastes-in-yarn-production
43. Abedin M, Siddique I. A research on reducing waste and better managing resources in the spinning industry using efficient methods. *Chemistry Research Journal.* 2024;9(2):40–9.
44. Ghosh SK, Gupta KR, Dey C. Study on generation of knotting waste in weaving cotton fabric and its remedial measures. *International Journal of Engineering Research & Technology (IJERT).* 2014 Jun;3(6):2185–94.
45. Daneshvar N, Ayazloo M, Khataee AR, Pourhassan M. Biological decolorization of dye solution containing Malachite Green by microalgae Cosmarium sp. *Bioresource Technology.* 2007 Apr;98(6):1176–82.
46. El-Rahim WMA. Assessment of textile dye remediation using biotic and abiotic agents. *The Journal of Basic Microbiology.* 2006 Aug;46(4):318–28.
47. Cripps C, Bumpus JA, Aust SD. Biodegradation of azo and heterocyclic dyes by Phanerochaete chrysosporium. *Applied and Environmental Microbiology.* 1990 Apr;56(4):1114–8.
48. Sekhar KC, Chary NS, Kamala CT, Vairamani M, Anjaneyulu Y, Balaram V, et al. Environmental risk assessment studies of heavy metal contamination in the industrial area of Kattedan, India—A case study. *Human and Ecological Risk Assessment: An International Journal.* 2006 Apr;12(2):408–22.
49. Banat IM, Nigam P, Singh D, Marchant R. Microbial decolorization of textile-dye-containing effluents: A review. *Bioresource Technology.* 1996 Dec;58(3):217–27.
50. Rissanen T. Inlaid fantasies: The textiles of Andreea Mandrescu. *Craft Res.* 2013 Oct 1;4(2):265–75.
51. Jordeva S, Tomovska E, Mojsov K, Golomeova-Longurova S, Maksimov S. Sustainability of the textile waste stream in Macedonia. *Advanced Technology.* 2018;7(1):74–8.
52. Haq UN, Alam SMR. Implementing circular economy principles in the apparel production process: Reusing pre-consumer waste for sustainability of environment and economy. *Clean Waste Systems.* 2023 Dec;6:100108.
53. Mwasiagi J, Mirembel J. Influence of spinning parameters on thin and thick places of rotor spun yarns. *International Journal of Computational and Experimental Science and Engineering.* 2018 Jul 2;4(2):1–7.
54. Lawrence CA. *Fundamentals of Spun Yarn Technology* [Internet]. Boca Raton, Florida: CRC Press; 2003 [cited 2023 Dec 1]. Available from: https://www.taylorfrancis.com/books/9780203009581

55. Gardetti MA and Torres AL. Sustainability in fashion and textiles—Values, design, production and consumption. *Management of Environmental Quality: An International Journal* [Internet]. 2013 Jun 7; pp-417-436 [cited 2023 Dec 1];24(4):417–436. Available from: https://www.emerald.com/insight/content/doi/10.1108/meq.2013.08324daa.012/full/html
56. Simegnaw Ahmmed A, Ayele M. In-depth analysis and defect reduction for Ethiopian cotton spinning industry based on TQM approach. *Journal of Engineering.* 2020 Apr 23;2020:1–8.
57. VLS Environmental Solutions. Hazardous vs. non-hazardous waste—What's the difference? [Internet]. [cited 2022 Jul 5]; Available from: https://www.vlses.com/2022/07/05/hazardous-vs-non-hazardous-waste/
58. Manshoven S, Mortensen LF, Smeets A, Tenhunen A. ETC/CE report 1/2022: Microplastic pollution from textile consumption in Europe. *European Topic Centre Circular Economy and Resource Use*, 2022.
59. Textile Value Chain. Digital media executive. Waste management methods in the textile industry. [Internet]. [cited 2022 Sep 12]; Available from: https://textilevaluechain.in/in-depth-analysis/articles/waste-management-methods-in-the-textile-industry
60. Reverse Resources. How much does garment industry actually waste? 2020 preferred fiber and materials market report. [Internet]. [cited 2021 Feb 1]; Available from: https://reverseresources.net/how-much-does-garment-industry-actually-waste/

3 Recycling Technologies and Challenges

3.1 INTRODUCTION

The textile industry utilizes more natural and non-renewable resources, leading to 10–20% of waste generation. Each year, nearly 2,000 million tons (MT) of municipal solid waste (MSW) are collected worldwide (1), with 65 to 92 MT of textile waste (2), and this is expected to reach 134 MT by 2030. To address this issue, diverse waste management systems have been implemented across various stages to minimize waste generation. The waste management hierarchy serves as a guiding framework for decision-making regarding waste management, both at individual and industrial scales (3). It categorizes methods into most and least preferred options, aiding in prioritizing actions. The waste management hierarchy, depicted in Figure 3.1, highlights preferred methods based on reusability, emphasizing strategies to minimize environmental impact.

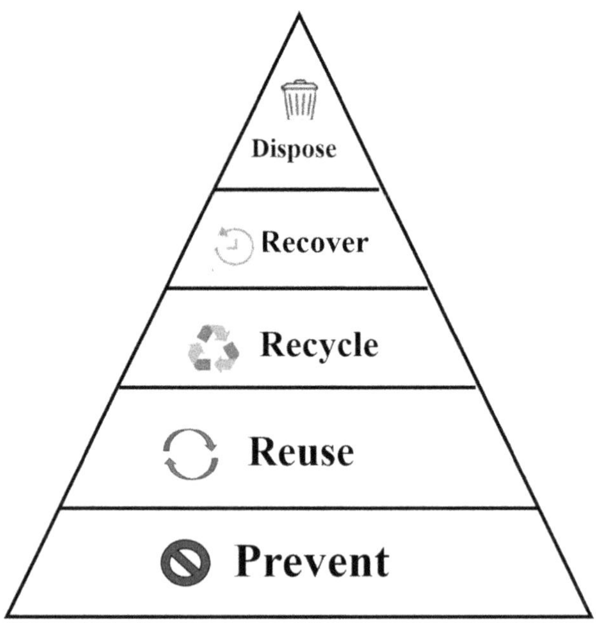

FIGURE 3.1 Waste management hierarchies.

DOI: 10.1201/9781003503835-3

3.1.1 Disposal

Disposal ranks as the least preferred waste management method due to its strain on land resources. In the European Union, about 10–20% of post-consumer textile waste is landfilled or incinerated. Polyester, the king of synthetic fabrics, makes up 55% of the global fibre market because of its low production cost and accessibility. Synthetic fibres' slow decomposition exacerbates the issue, alongside the release of toxic substances that pollute the atmosphere and natural resources. According to the Environmental Protection Agency, 18.6 MT of textile waste were disposed of in landfills, representing 5.8% of the total MSW generated that year. This is expected to increase to 150 MT by 2050. Generally, synthetic fibres take much longer to decompose than natural fibres, taking more than 200 years (4). Methane gas, significantly 28 times more harmful than carbon dioxide, is released as fabric decomposes, greatly impacting environmental well-being and being a major contributor to climate change (5). Human activities, including agriculture, fossil fuel operations, and waste management, are responsible for around two-thirds of methane emissions. This underscores the urgent need to mitigate disposal's environmental toll and transition towards more sustainable waste management practices.

3.1.2 Recovery

It involves generating electricity or heat from textile waste, commonly achieved through techniques like incineration, gasification, and anaerobic digestion. By 2030, global textile waste is projected to reach 148 MT, with over 35% consisting of cellulose-based materials. As a result, textile waste is increasingly seen as a valuable and sustainable resource for bioenergy production (6). Incineration entails combusting waste to produce ash, gas, and heat, particularly useful for treating specialized wastes like clinical or hazardous materials, reducing waste mass by 80–85% and volume by 95–96% (7). Gasification converts carbonaceous substances at high temperatures (>700°C) without combustion, yielding carbon monoxide, hydrogen, and carbon dioxide (8). Anaerobic digestion involves microbial breakdown of biodegradable materials in oxygen-deprived conditions, employed in industrial or domestic settings for waste management or fuel production. While incineration and gasification share similar process outlines, anaerobic digestion offers an alternative route for sustainable waste treatment and resource recovery.

3.1.3 Recycling

It involves converting waste materials into new ones, often including energy recovery. It is integral to waste reduction, forming the third component of the "Reduce, Reuse, Recycle" hierarchy. In India, only 59% of textile waste is recycled or reused, and only a small portion of that makes it back into the global supply chain. The remaining 41% is downcycled (19%), incinerated (5%), or ends up in a landfill (17%). The material composition of textile waste includes 60% made of cotton and cotton-rich materials and 19% made of synthetic-rich materials. Nearly 50% of energy is reduced by recycling synthetic fibres, and it also produces 79% less CO_2 emissions

FIGURE 3.2 Recycling, downcycling, and upcycling.

than virgin raw materials. Within recycling, materials undergo either downcycling or upcycling. Downcycling reduces material value and functionality but aids in resource conservation and pollution reduction. 2% of post-consumer waste is downcycled to make wipes and used for filling bedding, quilts, etc. Conversely, upcycling transforms waste into materials of higher value, aiming to maximize resource utilization and minimize waste. The illustration for recycling, downcycling, and upcycling symbols has been shown in Figure 3.2.

3.1.4 Reuse

It involves utilizing a product for its original purpose or repurposing it for a different use, aiding resource conservation and mitigating environmental impacts. Research demonstrates significant savings through reuse, with each kilogram of textile product reused instead of producing a new one saving 6,000 L of water, 3.6 kg of carbon dioxide, 0.3 kg of chemical fertilizer, and 0.2 kg of insecticides. 51% of domestic post-consumer textile waste is either recycled, resold, or exported to high-demand regions like Africa and Asia for use in second-hand clothing markets. In India, 16% of pre-consumer or post-industrial waste, resulting from overproduction or cutting waste, is repurposed for reuse in the manufacturing process (9).

3.1.5 Prevention

Prevention of waste encompasses actions implemented prior to substances, materials, or products becoming waste, aiming to diminish several factors. Firstly, it reduces waste volume by promoting product reuse and extending their lifespan. Secondly, it mitigates the detrimental effects of generated waste on both the environment and human health. Lastly, prevention strategies work towards minimizing the presence of harmful substances in materials and products, thereby reducing potential environmental and health risks.

3.1.6 Needs and Challenges of Recycling in the Textile Industry

The textile industry is a pivotal part of the global economy, generating significant trade, employment, and revenue. Global fibre production has skyrocketed in recent decades, growing from 70.6 MT in 2007 to 90.8 MT in 2014, with projections reaching 130 MT by 2025 (10). The textile sector uses 76 L of water each year, contributes to 20% of global industrial water pollution, and generates 35% of microplastic emissions (11). However, the sector faces growing challenges driven by overproduction and a culture of disposability, which exacerbate environmental degradation. To address these issues, textile recycling has emerged as a critical solution, serving as a pathway to a circular economy. Recycling offers the dual benefit of reducing environmental harm while aligning with sustainable business models. Yet, it is not without its obstacles, as represented in Figure 3.3. Interestingly, surveys have shown that most textiles can be recycled or repurposed. Around 60% of discarded clothes could be reused, 35% could be turned into wipes or recycled fibres, and only 5% need to be discarded (12). However, in reality, a large proportion of textile waste still ends up in landfills, where synthetic fibres, which dominate global production, fail to biodegrade.

3.1.7 Need for Recycling in the Textile Industry

3.1.7.1 Fast Fashion and Waste Reduction

The rise of fast fashion has led to a sharp increase in textile waste. As consumers demand affordable and trendy clothing, production volumes have surged, leading to high disposal rates. Recycling can help mitigate the negative impacts of fast fashion by reusing discarded textiles and preventing them from ending up in landfills.

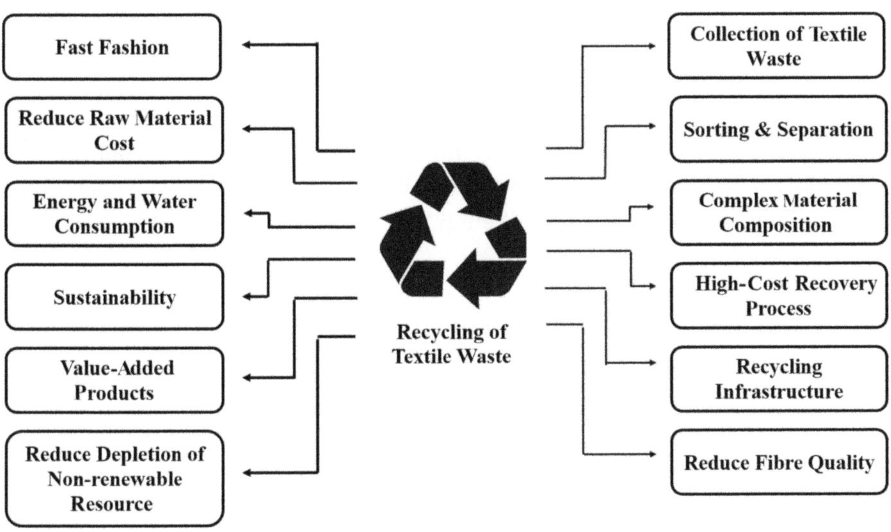

FIGURE 3.3 Needs and challenges of textile recycling.

3.1.7.2 Lowering Raw Material Costs

One of the primary incentives for textile recycling is the economic advantage it offers. Recycled materials are generally less expensive than virgin resources. This cost-effectiveness benefits manufacturers and significantly reduces the environmental costs associated with raw material extraction, such as deforestation and resource depletion.

3.1.7.3 Energy and Water Conservation

Recycling textiles can drastically reduce the consumption of energy and water. For instance, the recycling of post-consumer clothes cuts greenhouse gas emissions by 53%, decreases chemical pollution by 45%, and lowers water eutrophication—a harmful process that depletes aquatic oxygen—by 95% (11). Such savings are vital for an industry notorious for its resource-intensive processes.

3.1.7.4 Sustainability and Resource Preservation

Sustainability is no longer a choice—it is a responsibility. As the world's natural resources continue to shrink, there is a growing push for the textile industry to move towards circular methods. Recycling textiles plays a key role in this shift, allowing materials from discarded clothing and production waste to be reintroduced into the manufacturing cycle. This not only helps conserve non-renewable resources but also supports more responsible production practices.

3.1.7.5 Adding Value to Products

Generally, used textiles do not have to end up as waste—they can be given new life as value-added products. With growing interest in sustainability, many brands are now embracing recycled materials in their collections. This not only appeals to eco-minded customers but also sets these brands apart by showing they are serious about reducing their environmental footprint.

3.1.7.6 Reducing Non-renewable Resource Depletion

A large portion of textile fibres—around 63%—are made using petrochemical-based materials. If this situation is prolonged, carbon emissions could rise to about 35% by 2030, pushing levels over 400 MT. Recycling helps reduce the industry's dependence on these polluting sources, encouraging a move towards cleaner, more sustainable alternatives (13).

3.1.8 Challenges in Textile Recycling

There are several challenges that hinder the implementation of recycling textile waste on a large scale despite its potential. Some of these are as follows:

3.1.8.1 Collection and Sorting of Textile Waste

The biggest challenge is collecting and sorting textile waste efficiently because discarded garments are often made from different materials, which makes them challenging to separate for effective recycling. Sorting is also a labour-oriented process and not cost-effective. Textiles are often made from fibre blends like polyester/cotton or fabrics with spandex, buttons, zippers, etc. These combinations are hard to break down, and separating them requires either advanced machinery or chemical processes—both of which are expensive and technically complex.

3.1.8.2 High Recovery Costs

One of the biggest challenges in textile recycling is the high cost, particularly when dealing with blended or chemically treated fabrics. The expense involved in processing and converting waste into reusable materials often makes recycled products too costly—especially for small and medium-sized businesses.

3.1.8.3 Limited Recycling Infrastructure

In many areas, the infrastructure simply does not exist to support large-scale textile recycling. From basic collection systems to specialized recycling facilities, a lot of investment is needed to create a truly circular textile economy.

3.1.8.4 Declining Fibre Quality

Textiles lose strength and durability when recycled multiple times. The fibres tend to get shorter and weaker, which limits how they can be reused. To truly scale textile recycling, new methods are needed that can maintain or even improve fibre quality over time.

3.2 MECHANICAL RECYCLING OF TEXTILES

Mechanical recycling of textiles is a sustainable approach that reclaims used or discarded fabrics and transforms them into new fibres or products through physical processes without the intervention of any chemicals. This method is crucial for managing textile waste, conserving natural resources, and reducing the environmental impact of textile production. The process typically involves collecting, sorting, and processing used textiles (see Figure 3.4), which can include post-consumer waste from households and post-industrial waste from manufacturing. The mechanical recycling sequence includes tearing to reclaim fibres from used textiles to make new fabrics.

3.2.1 Collection

Textile waste is collected from households and industrial sources, typically managed by private companies and charities rather than municipalities. Globally, waste collection employs various service models encompassing two key elements: collection for disposal and collection for recovery. The former involves disposing of most textile waste in landfills or incinerators, costing the US approximately USD 4 billion annually in 2020. Charities have long facilitated textile reuse. In Sweden, for example, individuals donate 3 kg per person annually, amounting to 26,000 tons per year. Of this, 73% is exported as relief goods or sold abroad, 12% is sold domestically, and 15% is disposed of (14). Collaborative systems involving governments, municipalities, businesses, charities, and advocacy groups are essential for enhancing collection efficiency.

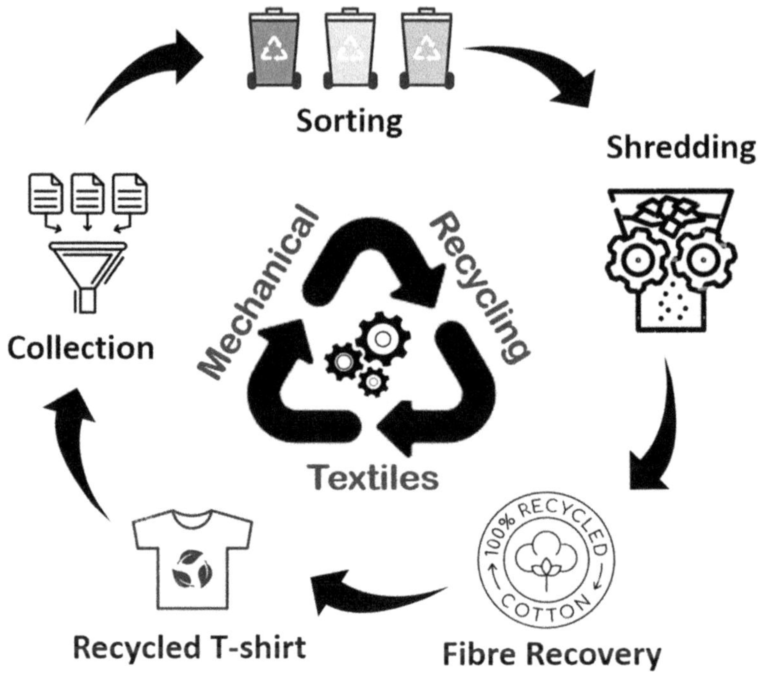

FIGURE 3.4 Mechanical recycling of textiles.

Textile waste accumulation in bins is irregular, with seasonal variations causing inefficiencies in collection frequency. Implementing modern technologies such as IoT and AI can improve efficiency by enabling real-time monitoring of bins, route optimization, and waste accumulation prediction. Smart bins integrated with sensor technology could reduce costs by 7.4% and CO_2 emissions by 10.2% compared to traditional systems (15). H&M, for instance, began using recycled materials in 2009 and launched its first Conscious Collection in 2010 (16). Across Europe, 15–25% of textile waste is collected separately, entering reuse or recycling streams. Reuse, focusing on high-quality reusable clothing, remains the priority, followed by recycling. These efforts underscore the need for innovative systems to improve textile waste collection and sustainability.

3.2.2 Sorting

The sorting and grading of collected textiles result in an inevitable fraction of low-grade waste unsuitable for reuse. Limited quantities of used textiles are collected and sorted for recycling due to challenges like fibre blends, which complicate sorting. Sorting primarily classifies textiles by fabric type, colour, and material composition to ensure high-quality recycled fibres. Waste is categorized into four types: fibre waste, cutting waste, fabric deadstock, and apparel overproduction. Smaller, unusable pieces are recycled, while larger reusable pieces are prioritized. High contamination levels and defects reduce acceptability, and materials like cotton are favoured for their recyclability. Sorting involves removing non-textile components such as zippers, buttons, and contaminants, a technique also applied in plastic recycling. Technologies used in textile sorting include:

1. **Manual sorting**: Workers classify textiles by touch, meeting recycler specifications for size, colour, and composition.
2. **Automated sorting**: Sensors identify fibre types based on light reflection, with compressed air directing textiles into the correct bins.
3. **Near-Infrared (NIR) technology**: NIR and visual spectroscopy (VIS) distinguish textiles by fibre type and colour. Technologies like Fibersort can process up to 900 kg of post-consumer textiles per hour using NIR spectroscopy.

At factories, fibre waste is segregated, while cutting waste is usually sold as mixed waste without sorting by size, composition, or colour. Some manufacturers separate white cuttings and larger panels due to high demand, fetching higher prices. White cutting waste sells at INR 50–60 per kg (USD 0.64–0.77), whereas mixed waste sells at INR 6–35 per kg (USD 0.07–0.45), depending on quality and market trends (9).

3.2.3 Shredding

After sorting and cleaning, the materials are compressed into bales for easier handling and feeding into recycling equipment. The first stage involves shredding the textile waste into fragments, followed by fibre recovery using a Garnett machine. This machine operates on the carding principle, aligning fibres and eliminating impurities to form a consistent web or batt of fibres. The Garnett machine employs a large rotating cylinder equipped with carding elements like wire teeth or rollers. These elements execute vigorous carding actions, tearing the fabric into its fundamental fibres. The teeth, mounted on parallel-axis cylinders, rotate in synchronized planes. While this effectively breaks down most textiles, yarns aligned with the motion of the teeth may pass through the machine without being fully broken into individual fibres. This shredding and carding process ensures that waste textiles are broken down efficiently into reusable fibres, forming a crucial step in textile recycling.

3.2.4 Fibre Recovery

Fibre separation is a crucial stage in textile recycling, restoring fabric into a reusable state. The shredding process significantly impacts the quality of recycled fibres, which are generally shorter due to the mechanical stress of tearing. Despite this, fibres exceeding 2 cm can be spun using open-end spinning techniques. Knitted fabrics, characterized by loosely spun fibres, achieve a recycling efficiency of about 75%, while woven fabrics face challenges due to tightly spun warp yarns. This leads to greater fibre breakage, producing lower-quality "shoddy" yarns. Recycling efficiency, reflecting usable fibre recovery, is influenced by fabric type. For a recycled polycotton yarn (45–55 blend) with a count of 28s, the CV% (coefficient of variation) is slightly higher (1.87) compared to the ideal 1.4. According to Reiter recycling classification, recycled fibres having less than 45% short fibre length (SFL) and 17 mm mean fibre length (MFL) are considered very good fibres, keeping cotton fibre of 24% short fibre content (SFC) and 21 mm MFL as reference (17). Woven fabrics from such yarns are recommended for suiting, while knitted fabrics of 140–145 gsm offer better potential. Additionally, recycled yarns cost 15–18% less than virgin fibre yarns, significantly reducing fabric production costs while promoting sustainable practices.

3.2.5 Advantages of Mechanical Recycling

Mechanical recycling of textiles brings both environmental and economic advantages. One major benefit is that it requires very little chemical processing because recycled fibres usually retain their original colours. This means the process does not use water and is much kinder to the environment. When coloured fabric is needed, white fabric waste is gathered, shredded, and dyed as necessary. The ability to mix recycled fibres with new ones gives manufacturers the flexibility to make a wide variety of fabrics for different uses. By cutting down on the need for virgin materials, mechanical recycling helps conserve natural resources like water and reduces reliance on petrochemicals. It also uses less energy than manufacturing textiles from raw materials, which means fewer greenhouse gases are released. In addition, it helps keep a large amount of textile waste out of landfills. For instance, recycling companies in India produce recycled yarns mostly in the 24s–28s count range, and they can create finer yarns by adjusting how recycled and new fibres are blended. The existing recycling infrastructure and affordable technology make mechanical recycling a practical choice for the industry.

3.2.6 Disadvantages of Mechanical Recycling

While mechanical recycling offers many benefits, it also comes with some challenges. Recycled fibres tend to be shorter and often do not match the quality of new, virgin fibres, which limits their use in high-end products. Fabrics made from blends, as well as non-textile parts like buttons and zippers, make the recycling process more complicated and can reduce the quality of the final material. Additionally, sorting, cleaning, and processing textile waste can be costly, especially when dealing with synthetic fibre blends, which are tougher to recycle mechanically. Overcoming these issues will require improved technologies for sorting and cleaning, along with new recycling methods. Cooperation across the industry and clear standards will be crucial to make mechanical recycling more practical and widely accepted.

Mechanical recycling can work in two main ways: closed-loop and open-loop systems. In closed-loop recycling, synthetic fibres are melted down to create new fibres, but this method needs pure, uncontaminated materials and cannot handle certain fibre blends or polymers that do not melt easily. The shredding method used in open-loop recycling removes any metal or plastic parts, cuts the textiles into small pieces, and then feeds them into machines that separate the fibres. These fibres are then carded and spun into yarn. Mechanical recycling generally preserves the fibre's properties except for length, and by sorting fabrics by colour, it can often avoid the need for re-dyeing, which helps reduce the environmental footprint.

3.3 THERMAL RECYCLING OF SYNTHETIC TEXTILES

The production of synthetic textiles has risen sharply due to the growth of fast fashion. This has created major challenges for traditional recycling methods because synthetic fabrics are made from complex polymers derived from petrochemicals. Tackling these issues is vital for making the industry more sustainable. One promising solution is thermal recycling—also called mechanical polymer recycling. This technique involves melting synthetic fibres so they can be re-spun into new fibres or reshaped into new materials. Thermal recycling breaks synthetic polymers back into their basic building blocks, which can then be reused in manufacturing. It works especially well with materials like Nylon 6 and thermoplastic polyurethane. However, widespread use is limited due to the high costs involved in removing contaminants. Despite this, companies like Teijin in Japan have successfully demonstrated that thermal recycling can be a practical option.

Thermo-mechanical recycling is especially efficient for pre-consumer waste, such as production scraps, where material consistency simplifies processing. Through mechanical shredding, waste is converted into chips or pellets, which are then subjected to melt extrusion—a high-temperature process that reshapes the molten polymer into new fibres (see Figure 3.5). This approach is versatile and cost-effective, offering significant advantages for recycling common materials like polyester, nylon, elastane, polypropylene, and polyurethane.

The versatility of thermal recycling enables it to address a wide range of synthetic and natural fibres, including cotton, polyester, nylon, and other blends. By integrating **thermo-chemical recycling** technologies such as pyrolysis and gasification, the process contributes to a circular economy, transforming textile waste into valuable raw materials while reducing environmental impact (18). While challenges such as decontamination costs and scaling remain, the potential of thermal recycling to revolutionize the management of synthetic textile waste is undeniable, offering a pathway to a more sustainable future for the fashion industry.

Pyrolysis involves heating materials at high temperatures, such as synthetic textiles, in the absence of oxygen. This process breaks down polymer chains into byproducts such as char, oil (a petrochemical liquid), and syngas, which is a mixture of gases, as illustrated in Figure 3.6. Syngas—composed mainly of carbon monoxide

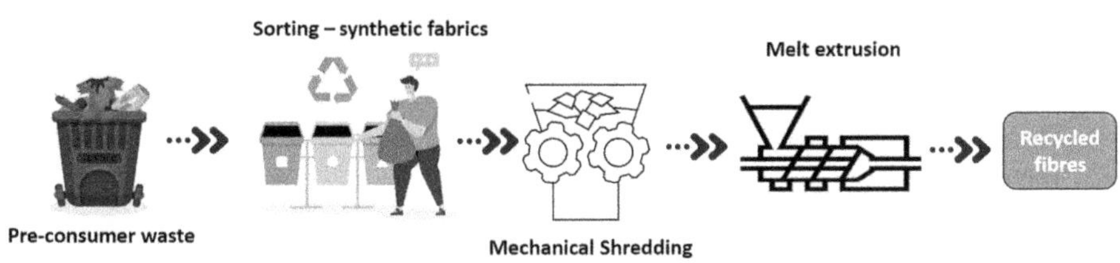

FIGURE 3.5 Thermo-mechanical recycling process.

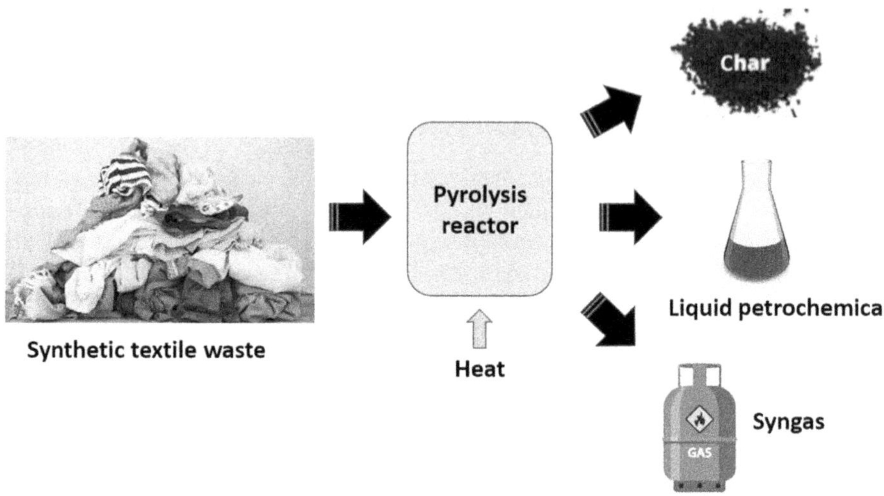

FIGURE 3.6 Pyrolysis process.

and hydrogen—can be reused to produce chemicals or fuels. These by-products serve as valuable raw materials for chemical manufacturing (19). Pyrolysis is particularly effective for handling complex polymer-based waste.

Gasification is a process in which organic materials are partially burnt at very high temperatures. The by-product of this process is syngas, a mixture of carbon monoxide, carbon dioxide, and hydrogen. Syngas can be used as a fuel or as a basic ingredient to make chemicals. This makes gasification a flexible way to convert waste into energy or new products (see Figure 3.7).

3.3.1 Advantages and Disadvantages of Thermal Recycling

Thermal recycling has specific benefits for synthetic textiles. Unlike mechanical recycling, it can process a wide variety of synthetic fabrics, even blends that are usually challenging to recycle. This makes thermal recycling a more flexible and better option for managing different types of textile waste. The process creates high-quality materials that can be reused in the textile industry, which means the use of materials

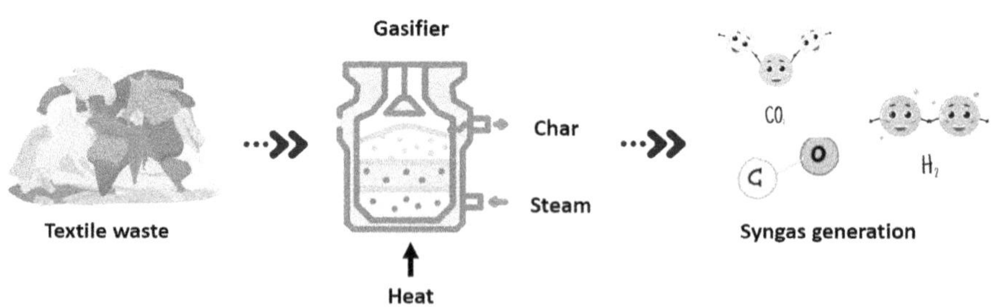

FIGURE 3.7 Gasification process.

from fossil fuels may be reduced. By turning old textiles into reusable resources, thermal recycling helps reduce the environmental impact of textile production and supports sustainability. Synthetic fibres do not break down easily using conventional recycling methods; hence, efficient thermal recycling processes pave the way for research and innovation. It is also important to note that textile waste undergoing thermal recycling should be free from dirt and other impurities.

Synthetic textiles are used in most applications, from clothing and furniture to industrial fabrics. Setting up a large-scale thermal recycling manufacturing unit is not economically easy; it requires space, advanced equipment, and trained professionals to operate the industry. Moreover, collection and sorting systems are crucial to ensure synthetic textiles are separated from other materials and cleaned for processing. While these challenges are significant, the recycling process can be made more efficient and stable. There are many growing companies and research groups focusing on improving thermal recycling methods to make them more cost-effective, scalable, and eco-friendly so that recycling synthetic textile waste becomes easier.

3.4 CHEMICAL RECYCLING OF NATURAL FIBRES

Chemical recycling of natural fibres paves the way to handle textile waste by turning waste fibres like cotton back into regenerated fibres, which can be used to make new textiles. This process uses chemicals to break down the fabric at the molecular level, converting the fibres into high-quality fibre better than virgin fibres. Compared to mechanical recycling, chemical recycling is more flexible and can be used to develop a wider range of products. It can be incorporated with different types of fabrics and can be used to develop products from jackets to car panels and home décor.

This method focuses on breaking down the polymers of natural fibres such as cotton, wool, and silk into their original monomers. Unlike mechanical recycling, which can reduce the quality of fibres, chemical recycling recovers key components like cellulose, which can then be used to produce new fibres or other products. There are several techniques involved in this process, such as pulping, hydrolysis, solvolysis, and pyrolysis, which turn waste textiles back into valuable raw materials.

3.4.1 Pulping Process

The pulping process is one of the major techniques for recycling cellulosic fibres like cotton through chemicals. Instead of breaking cellulose down completely into its basic building blocks, this method partially breaks the cellulose chains. The material produced can then be used to make regenerated cellulosic fibres—such as viscose, Lyocell, and Ioncell-F—using solution spinning methods. In this process, cellulose-rich waste like cotton textiles is converted into a usable pulp, as illustrated in Figure 3.8. This pulp can then be spun into new fibres for textile production.

The pulping process involves suspending cotton cellulose in liquid containing specific chemicals, leading to its partial depolymerization. This results in slurry or dissolving pulp. The viscosity and reactivity of the cellulose are adjusted, often through chemical treatments designed to remove dyes and finishes, followed by

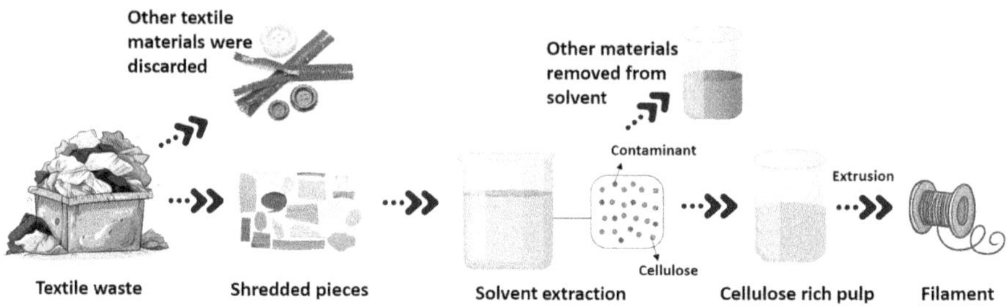

FIGURE 3.8 Pulping process.

bleaching. Various environmentally friendly chemicals used in closed-loop systems facilitate the spinning of regenerated cellulose fibres, such as Lyocell, carbamate, and Ioncell.

Examples of the pulping process include the development of viscose fibre from cotton pulp. This method involves depolymerization of 100% cotton fabric into pulp. This pulp is then processed into cellulosic fibres, similar to how viscose is produced from wood pulp. However, the resulting polymer chains from recycled cotton are weaker than those derived from wood pulp, leading to reduced strength. To mitigate this, regenerated fibres are often blended with virgin viscose fibres.

Most technologies for recycling pure cotton textiles through the pulping process have achieved a high Technology Readiness Level (TRL) of 7 to 9. It is expected that these technologies will reach TRL 9 by 2025. Current process capacities vary significantly, from as little as 10 kg per day to thousands of tons annually. However, scaling up these processes further requires more production time, customer feedback for optimization, and a continuous supply of suitable textile waste in terms of purity and composition. Although some technology developers are exploring the recycling of polycotton blends, this approach is less economically attractive due to the additional steps required for separating and purifying the fibres.

3.4.2 Hydrolysis

It involves breaking down natural fibres into their constituent components, such as glucose, cellulose, hemicellulose, and lignin, using water combined with acids or bases. Hydrolysis is effective but requires careful control of reaction conditions to maximize yield and purity. For example, cotton waste textiles are broken down through depolymerization to produce a glucose solution, using sulfuric acid as the only catalyst, as shown in Figure 3.9. This method achieves a high yield of over 70%, and the resulting glucose solution has a high concentration exceeding 100 g/L. The high concentration reduces purification costs, enhancing the process's industrial viability. The glucose produced can then be fermented into butanediol or caprolactam, which are key precursors for manufacturing synthetic textile fibres. This approach helps retain the value of waste textiles within the textile production cycle. In the **solvolysis** process, solvents like alcohols or esters dissolve the natural fibres

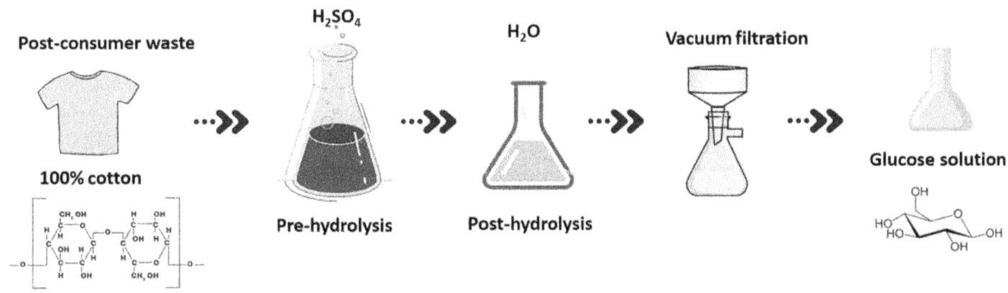

FIGURE 3.9 Acid hydrolysis process.

and separate them into their original chemical components. Another method involves enzyme hydrolysis using a bioreactor for processing cotton. This approach dissolves the cotton fibres, leaving the polyester fibres intact, providing a clear path for their separate recovery and reuse.

3.4.3 Methanolysis

Methanolysis offers an efficient way to repurpose cellulose fibres, enhancing the sustainability of textile recycling. Large quantities of terry towels are used in hotels and hospitality industries, generating significant cellulosic waste upon disposal. Textile recycling, whether "cradle to grave" or "cradle to cradle," has gained momentum recently. Unlike dyed terry towels, white terry towels, even when soiled, do not require dye stripping, making them an excellent source of waste cellulose for Carboxyl Methyl Cellulose (CMC) conversion used in textile printing, as shown in Figure 3.10.

Pure white cellulosic terry towels are sourced, cut into tiny fibrous masses, and suspended in isopropyl alcohol. Mechanical stirring applied, and 40% NaOH is added to the mixture, which is stirred for 90 min to obtain alkali cellulose. Monochloroacetic acid is then slowly added over 30 min, and the mixture is maintained at 55°C for 3.5 h. Methanol (70% v/v) is subsequently added, and the mixture is neutralized with acetic acid (90% v/v). CMC is recovered via filtration and washed six times with ethanol/water (70/30 v/v). The product is further washed with pure methanol and oven-dried at 60°C. To increase the degree of substitution (DS), synthesized CMC underwent a second etherification reaction. The resulting CMC is ground, filtered through a 60-mesh nylon cloth, and stored in an airtight container for textile printing paste preparation.

The study indicates that fabrics printed using the developed CMC exhibit higher K/S values compared to those printed using conventional methods. Blends of these two thickener pastes resulted in marginally increased K/S values with higher proportions of laboratory-made CMC paste in the blend. Rheology studies revealed the pseudo-plastic and shear thinning behaviour of the prepared thickener, consistent with CMC's properties.

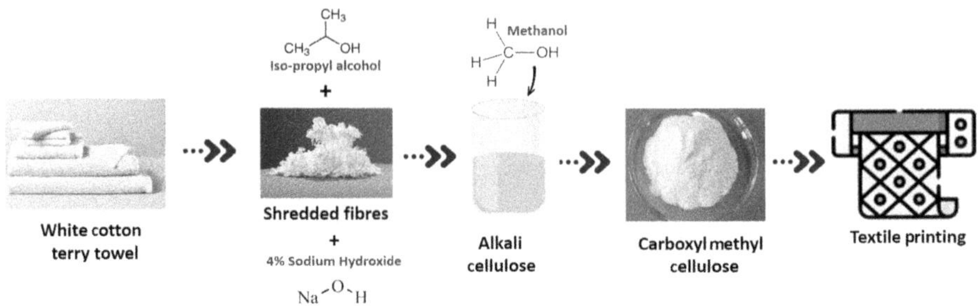

FIGURE 3.10 Methanolysis process.

3.4.4 Pyrolysis

It decomposes natural fibres into their chemical components by heating them in the absence of oxygen. This method is more commonly associated with the treatment of synthetic textiles but can also be applied to natural fibres under specific conditions. Another notable method of chemical recycling involves the separation of fibre constituents in polycotton blends. This is particularly valuable because most fabrics contain blended fibres, making mechanical separation challenging. HKRITA (Hong Kong Research Institute of Textiles and Apparel) has developed a technology known as The Green Machine, which uses a solvent-based process to separate polyester fibres from polycotton blends. This method uses heat, water, and a small amount—less than 5%—of biodegradable green chemicals to break down the cotton portion of blended fabrics into liquid cellulose. After drying, the process yields polyester fibres with their original quality intact, along with a usable cellulose powder.

3.4.5 Challenges in Chemical Recycling of Natural Fibres

Natural fibres often come with a mixture of cellulose, hemicellulose, lignin, and other impurities. Efficiently separating these components to isolate pure cellulose is challenging, yet essential for producing high-quality recycled materials. Many chemical recycling methods are energy-intensive and use harsh chemicals, raising concerns about their overall sustainability. Developing cost-effective and environmentally friendly technologies is critical for broader adoption. Scaling chemical recycling processes to handle large volumes of natural fibres and ensuring the quality and purity of the recycled output are significant hurdles. Industry standards and certifications are needed to ensure consistent quality and build consumer trust.

The challenge with recycling different sources of cellulose, like wood and cotton, lies in their varying chemical structures and viscosities, which necessitate tailored adaptations in the pulping process or pre-treatment procedures. For the pulping process to be economically viable, the input material must contain at least 50% cotton. Lower concentrations are less feasible due to the inefficiency and higher costs associated with processing. Some technologies focus on separating polyethylene terephthalate (PET) from cotton in blends through solvent extraction, but the primary emphasis remains on recycling the cotton component. Unlike mechanical recycling, which can be affected by the structural integrity of the fabric, the chemical pulping process can handle diverse textile forms, though dyed textiles may require additional decolouring or bleaching steps. The efficiency of the pulping process is highly contingent on the purity of the input material. Both pre-consumer and post-consumer textile waste can be processed, but contamination and impurities significantly reduce yield and increase operational costs.

3.4.6 Advancements in Chemical Recycling

While innovative technologies like NuCycl, Circulose®, Ioncell, Refibra, Saxcell, and Infinited Fiber lead to significant potential, they also face challenges such as high costs, energy use, and chemical reliance (20).

NuCycl is a process that transforms waste textiles into a new fibre that can be converted into various applications. The process involves collecting and shredding the textile waste, converting it into a purified pulp using sustainable chemicals, and then spinning it into Lyocell fibres. The recycled fibres are strong, fine, and have similar properties to virgin fibres, which makes them suitable for clothing, shoes, and home products. However, the technology is still not cost-effective or energy-saving and depends on chemicals, which hinders its overall sustainability.

The Renewcell process produces a fibre called Circulose from discarded cotton materials and other natural fibres. Generally, this method dissolves the textile waste in a chemical, removes the impurities, and dries the filtrate into pulp sheets. Further, these sheets are converted into regenerated materials like viscose or Lyocell. Circulose is made completely from textile waste, and hence it is a biodegradable and more sustainable alternative to virgin fibres. Still, the process involves chemical use and high pre-treatment costs, which can limit both its environmental and financial appeal.

Ioncell offers an eco-friendlier process by using non-toxic ionic liquids to dissolve cellulose from textile waste. The resulting fibres are soft like silk, reflect colour well, and are mechanically strong. It avoids harmful co-solvents, making it safer for the environment. Despite these benefits, Ioncell is still in the development stage and faces high energy requirements and costly pre-treatment steps that make scaling up difficult.

Refibra technology blends cotton scraps with responsibly sourced wood pulp to create TENCEL™ Lyocell fibres. This process happens in a closed-loop system that reuses a safe solvent, reducing environmental harm and aligning well with circular economy goals. It is already being used on a larger scale, but like others, it comes with high costs and relies on chemicals, especially during the initial stages of processing.

The use of advanced solvents, like ionic liquids and supercritical fluids, can enhance the efficiency and sustainability of chemical recycling processes. These solvents can dissolve cellulose more effectively, simplifying the separation of pure cellulose from other components. Techniques such as membrane processes and chromatography can improve the purity of recycled cellulose and other by-products. These technologies are crucial for producing high-quality materials that meet industry standards (21).

3.5 DEPOLYMERIZATION AND RECYCLING

Depolymerization and recycling processes are essential in converting polymer-based materials back into their constituent monomers for reuse, contributing significantly to sustainability efforts. This process involves chemically sorting textile waste by fibre type, depolymerizing the polymers into their monomers, and repolymerizing them to create new polymers. The recycled polymers retain properties identical to virgin polymers, enabling their conversion into fibres, yarns, fabrics, and ultimately garments, which can be recycled again (see Figure 3.11). This approach addresses plastic pollution, reduces dependence on fossil fuels, and supports a circular economy (22).

3.5.1 Depolymerization of Nylon 6 to Caprolactam

Numerous depolymerization methods are employed for Nylon 6 recycling (23), including hydrolysis, ammonolysis, acid-catalyzed, base-catalyzed, water-based, and alcoholysis processes (see Figure 3.12).

3.5.1.1 Hydrolysis

Hydrolysis can be categorized into acidic, neutral, and alkaline. Acidic hydrolysis uses agents like phosphoric acid (H_3PO_4) or hydrochloric acid (HCl). Typical reaction conditions are moderate temperatures (200–300°C). Acidic hydrolysis produces ε-caprolactam and aminocaproic acid (ACA), but substantial by-products (e.g., ACA

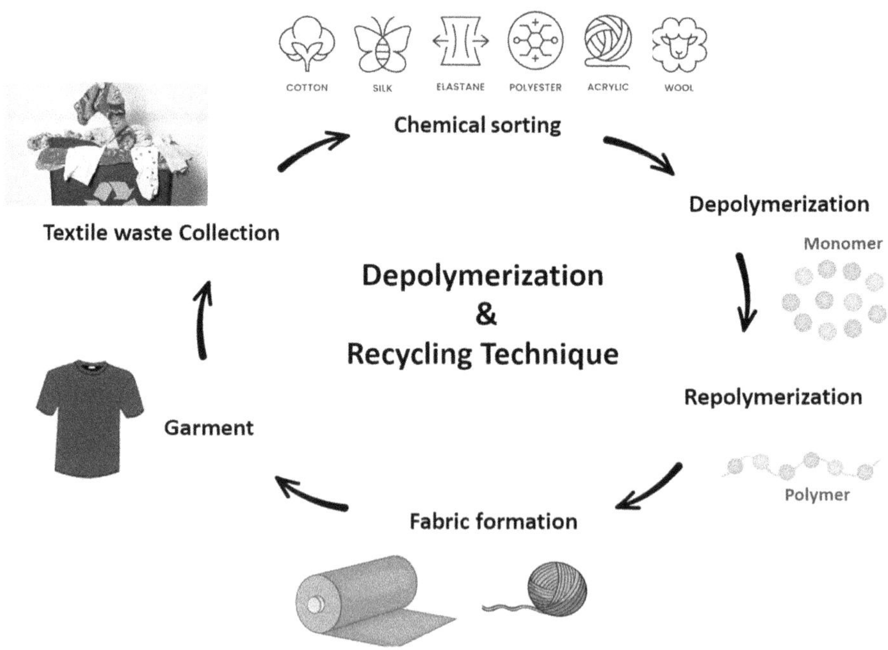

FIGURE 3.11 Depolymerization and recycling techniques.

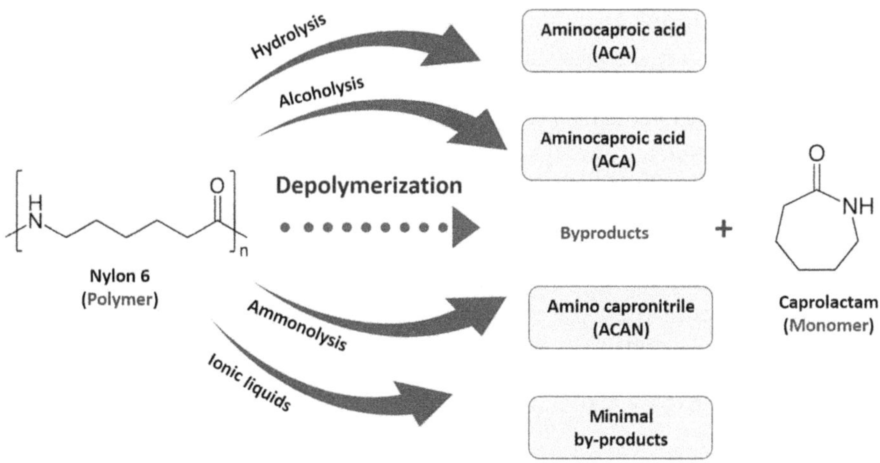

FIGURE 3.12 Depolymerization of Nylon 6 to caprolactam.

salts) can complicate the recovery of caprolactam. Neutral hydrolysis employs high-temperature water (300–400°C) and often results in simpler reactions but requires energy-intensive separation due to by-products. Alkaline hydrolysis, on the other hand, uses sodium hydroxide (NaOH) under vacuum, providing selective conversion to caprolactam without significant by-products. Among these, alkaline hydrolysis is highly energy-efficient and selective.

3.5.1.2 Ammonolysis

Ammonolysis involves ammonia to break Nylon 6 bonds at high temperature (300–350°C) and pressure (~140 bar). This method primarily produces aminocapronitrile (ACAN) alongside ε-caprolactam. However, side reactions lead to ACAN conversion to other derivatives, posing purification challenges. Ammonolysis is rarely employed for caprolactam recovery due to its limited yield and challenging processing conditions.

3.5.1.3 Alcoholysis

Alcoholysis utilizes alcohols like isopropanol (iPrOH) or methanol under subcritical or supercritical conditions. A temperature of 300–370°C and high pressure (~20 MPa) ensures the breakdown of Nylon 6 into caprolactam. This method shows reduced by-products when secondary alcohols are used but can lead to N-alkylation of caprolactam with primary alcohols. Alcoholysis is less energy-efficient but can recover solvents effectively for reuse.

3.5.1.4 Using Ionic Liquids (ILs)

Ionic liquids such as 1-butyl-3-methylimidazolium chloride or tetrafluoroborate operate under relatively lower temperatures (150–300°C). ILs enable a selective back-biting depolymerization mechanism, providing high-purity caprolactam with minimal by-products. However, their cost and difficulty in recycling remain barriers to large-scale use.

3.5.1.5 Ring-Closing Depolymerization

This process uses organo-catalysts and reagents to break down the Nylon 6 polymer into low molecular weight chemicals, which can then be converted back into monomers for new production cycles. Alberti et al. conducted a study on the chemical recycling of Nylon 6 through a ring-closing depolymerization process (24). In their research, depolymerization of Nylon 6 is achieved using DMAP (dimethylaminopyridine) as an organo-catalyst and acetic anhydride as a depolymerization reagent. During the process, Nylon 6 is subjected to microwave heating at 250°C for 15 min, yielding low molecular weight chemical N-acetylcaprolactam in good yields, despite the relatively low reaction temperature. The depolymerization process produces N-acetylcaprolactam, which must undergo deacetylation to convert it into ε-caprolactam—a key raw material used to make new Nylon 6 (see Figure 3.13). This step closes the recycling loop, enabling the production of Nylon 6 without any loss in quality.

3.5.2 Depolymerization of PET Waste into Monomers

Glycolysis: This method breaks down PET waste into the monomer bis(2-hydroxyethyl) terephthalate (BHET) using ethylene glycol as the depolymerizing agent and zinc acetate as the catalyst (see Figure 3.14). A 5:1 weight ratio of ethylene glycol to PET is used, and the mixture is placed in a reaction vessel along with zinc acetate, added in proportion to the weight of the PET waste. Hot distilled water is added to each setup to remove excess ethylene glycol and zinc catalyst, followed by filtration. The residual part is collected, dried, and cooled to obtain white crystalline solids. These obtained monomers serve as raw materials for producing PET-based products (25).

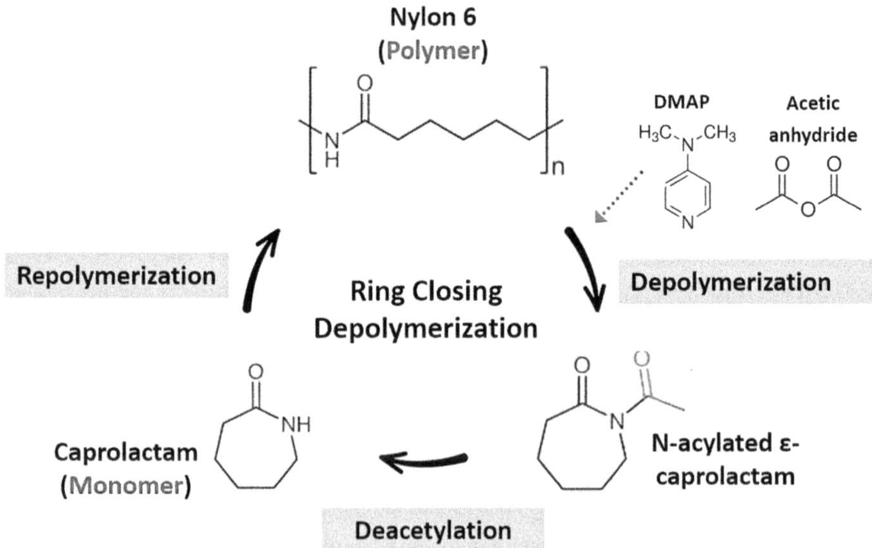

FIGURE 3.13 Ring-closing depolymerization.

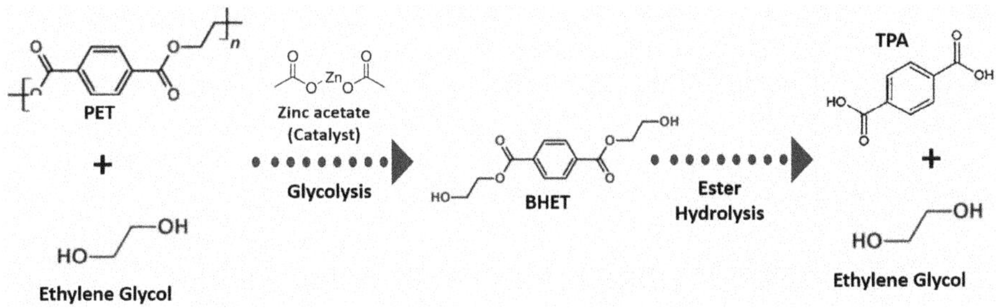

FIGURE 3.14 Depolymerization of PET into monomer by glycolysis and hydrolysis process.

This process can reclaim high-quality PET from post-consumer recycled clothing, providing a valuable feedstock for new products. The ester hydrolysis process is used to break down BHET into the true monomers of PET: terephthalic acid (TPA) and ethylene glycol (26).

3.5.2.1 Enzymatic Recycling

In plastic waste management, enzymatic recycling of PET has emerged as a sustainable approach (27). This method uses PET-depolymerase enzymes to catalyze the depolymerization of PET under mild conditions, compared to the high temperatures and harsh chemicals typically required in conventional chemical recycling. Recent advancements have led to the development of highly efficient engineered PET-depolymerase enzymes capable of hydrolyzing the ester bonds in PET polymers. These enzymes break PET down into monomers, namely, TPA and ethylene glycol (EG) (28). The biocatalytic process works under ambient temperatures and neutral pH, which significantly reduces the energy and environmental footprint associated with PET recycling.

The enzymatic recycling method is flexible, capable of breaking down any forms of PET, even when they contain common additives or contaminants. The enzyme's nature makes it adaptable to interact directly with polymer chains at the molecular level, eliminating the need for additional pre-treatment. Recent advances in protein engineering and directed evolution have significantly improved the enzyme's performance—boosting its activity, stability, and ability to target PET specifically. Further progress in reaction kinetics and process optimization has led to highly efficient depolymerization, enabling the recovery of PET monomers with near-complete yield and high purity. The recovered monomers are repolymerized into new PET, closing the loop of the recycling system. As a result, enzymatic recycling offers a promising and sustainable solution to the global PET waste problem, supporting circular economy efforts and reducing reliance on fossil fuel-based raw materials.

3.5.3 Challenges in Depolymerization

The presence of polymers, additives, and blends makes it difficult to develop standard depolymerization processes because each type of polymer may require a specific method for effective breakdown and recovery. Textile materials often contain dyes, fillers, and other additives that make it challenging to choose the appropriate depolymerization process. Removing these contaminants is essential for producing high-quality recycled materials. The depolymerization processes can be energy-intensive and costly, which makes them less attractive compared to producing plastics from virgin materials. There is a lack of awareness about the regulations and standards for recycled materials, which causes differences in quality and safety.

3.6 RECYCLING OF FIBRE-REINFORCED COMPOSITES

Fibre-reinforced composites (FRCs) mean incorporating fibres (CF) as a reinforcement in a composite plays a significant role in technical textile industries such as aerospace, automotive, and renewable energy due to their high strength-to-weight ratios. Carbon fibres have excellent mechanical properties, offering advantages in producing lighter, stronger, and more durable materials. The global carbon fibre market was valued at USD 7.48 billion in 2023 and is expected to grow at a compound annual growth rate (CAGR) of 12.9% from 2024 to 2032. However, the prolonged use of FRCs leads to non-biodegradable waste, and hence managing end-of-life for those composite wastes has become a critical environmental and economic challenge.

The wind turbine industry alone is projected to generate approximately 483,000 tons of carbon fibre-reinforced composite (CFRC) waste by 2050 in Europe (29). This growing waste has become a major problem due to its high cost for recycling as well as the environmental impacts of producing virgin carbon fibres (vCF). The manufacturing process for vCF is both energy-intensive and produces harmful by-products. Consequently, there is a pressing need to develop recycling methods for reusing waste carbon fibres to mitigate these impacts (30).

3.6.1 Mechanical Recycling

Mechanical recycling is one of the primary methods explored in research for managing CF waste (31). This process involves reducing the composite materials into smaller parts through shredding, crushing, grinding, and milling. There is specialized equipment, such as multiple-shaft shredders and cutting mills, generally used to convert the small parts into uniform flakes, controlled by blade distances, sieve sizes, and rotation speeds. After shredding, these flakes can be ground into fine powders. These powders are mixed with new materials and reformed into new composites. For example, carbon fibre-reinforced polyphenylene sulphide composites are mechanically processed into flakes, which are then blended with new fillers and reshaped through low shear mixing and compression moulding. This process helps to maintain the integrity of fibre strength and quality.

3.6.2 Thermal Recycling

Pyrolysis is one of the thermal recycling processes that reclaims the carbon fibres from waste composite materials (32). Pyrolysis involves decomposing the organic matrix in FRCs by heating in an inert atmosphere, such as vacuum, superheated steam, or nitrogen gas. This process breaks down the polymer matrix and is cleaned to achieve virgin-quality carbon fibres. A crucial step involved in the pyrolysis process is optimization of temperature and duration so that the carbon fibres are not degraded during the process. Another thermal method includes the fluidized-bed process which utilizes a high flow of gas to separate the composite particles by balancing the gravitational and upward forces simultaneously (33). This method decomposes the matrix component at elevated temperatures and liberates the carbon fibres without any significant damage. Hence, selecting the optimal processing temperatures is important during the process to prevent fibre degradation while ensuring efficient matrix decomposition.

3.6.3 Chemical Recycling

Chemical recycling involves breaking down the polymer matrix in FRCs through the use of solvents or chemical treatments, leaving the reinforcing fibres intact. The nitric acid recycling method could decompose the epoxy resin contained in CFRPs under relatively moderate conditions (80°C and atmospheric pressure) (34). This process can effectively recover high-quality fibres from composite waste, which can be reused in new composite materials. Chemical recycling methods must address the strong adhesion between fibres and the polymer matrix, as well as the presence of contaminants like fillers, pigments, or coatings, which can complicate the recycling process.

3.6.4 Challenges in Recycling Fibre-Reinforced Composite

- The strong bond between reinforcing fibres and the polymer matrix makes separation difficult without damaging the fibres. Effective separation techniques are essential for recovering high-quality fibres while maintaining their mechanical properties.
- During mechanical recycling, processes like grinding and shredding can cause fibre breakage and reduce the strength and stiffness of the recycled fibres. Hence, reducing fibre degradation is a crucial factor for ensuring the mechanical properties of recycled composite materials.
- The technologies developed for recycling FRCs are still in research and not yet commercialized due to their high cost, lack of infrastructure, and requirement for advanced technologies.
- Standard regulations need to be framed to ensure the performance and durability of recycled composites, leading to acceptance across various industries.

Recycling FRCs is key to reducing their environmental footprint and preserving valuable resources. As industries increasingly rely on lightweight, high-strength materials, finding sustainable ways to manage FRCs at the end of their life cycle is becoming more critical. Future efforts should aim to make recycling methods more efficient and affordable. This means refining mechanical recycling to protect the fibres from damage, fine-tuning thermal processes to recover better-quality fibres, and advancing chemical techniques to separate fibres from complex resin systems more effectively. It is also important to create clear standards for recycled FRCs to help encourage their use in new products. Tackling these issues will move the composite materials industry towards a more sustainable future and support a circular economy where materials are reused instead of wasted.

3.7 DYES AND FINISHING AGENTS AND THEIR IMPACTS ON RECYCLING

Dyes and finishing agents are integral to the textile industry, providing fabrics with vibrant colours, unique patterns, and enhanced properties like durability and water resistance. However, as the industry faces growing pressures to adopt sustainable practices, the role of these chemicals in textile recycling has come under increased scrutiny. With millions of tons of textile waste produced globally each year, effective recycling strategies are critical for reducing environmental impact and conserving resources.

3.7.1 Impacts of Dyed and Finished Textiles on Environment

Dyed and finished textiles have a substantial environmental impact and their manufacture, use, and disposal present a number of interrelated problems (see Figure 3.15). The challenge of recycling these materials for future use is a significant obstacle. Sorting, a crucial recycling step, is made more difficult by the heterogeneous fabric mixes found in dyed and finished textiles. The ability to separate materials by type and quality is the foundation of traditional recycling techniques, but the addition of dyes and finishes adds more layers of complexity. Sorting becomes inefficient, expensive, and labour-intensive which impedes extensive recycling initiatives. Additionally, decolourization procedures are needed to remove colours and finishes,

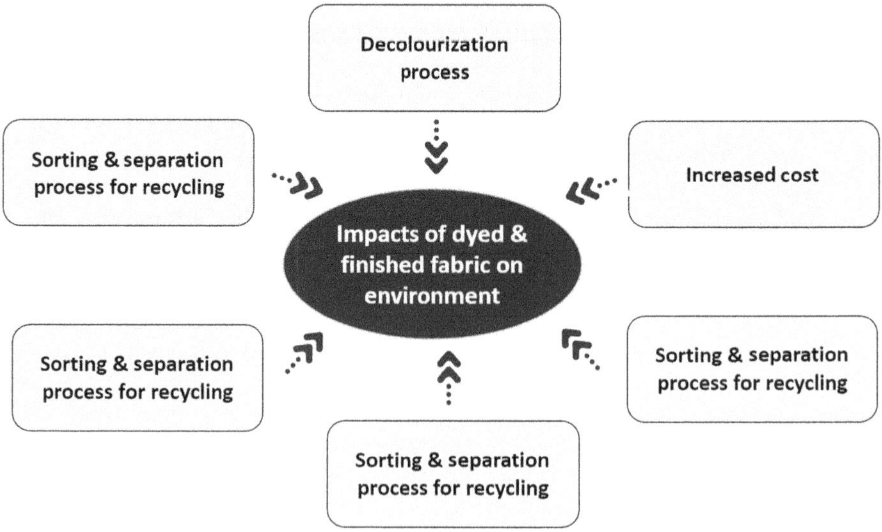

FIGURE 3.15 Impacts of dyed and finished fabric on environment.

which add to the environmental burden by using chemicals and energy. This difficulty emphasizes the need for technological advancements in textile recycling.

Additionally, dangerous chemicals are added to the cycles of production and recycling through the use of dyes and finishing agents. Hazardous substances like formaldehyde, azo compounds, and heavy metals are present in a lot of synthetic dyes and finishing agents. When produced or disposed of, these chemicals have the potential to leak into waterways, endangering aquatic life, and contaminating ecosystems. Consumers and employees may be at risk of health problems if these dangerous chemicals contaminate new products or stay in the fibres after recycling. Additionally, managing dyed and finished textiles comes with higher financial and environmental costs due to the need for additional resources for the safe handling and treatment of these chemicals.

Pollution and wastewater production during the textile production and dyeing processes have a major influence as well. One of the most water-intensive sectors of the economy is textile dyeing, and the wastewater from dyeing facilities is frequently filled with hazardous dyes, salts, and other components. When wastewater is not properly treated, pollutants may be released into rivers and oceans, seriously harming the environment, and having an impact on nearby communities. Since synthetic dyes are difficult to remove from the environment, soil and water become contaminated over time.

In addition to having an impact on ecosystems, this pollution lowers the amount of clean water available for agricultural and human use. In addition to pollution, the depletion of non-renewable resources caused by the production of dyes and finishing agents also contributes to environmental degradation. Numerous of these compounds come from petrochemicals, which have a significant carbon footprint during their extraction and processing. As the demand for textiles rises worldwide, sustainability issues are made worse by the dependence on limited resources. Moreover, hazardous materials are released into the environment during the disposal of textiles containing these agents, whether by landfilling or incineration. While landfilling causes chemicals to slowly seep into the groundwater and soil, incineration releases harmful gases.

Last but not least, using dyed and completed textiles raises production and waste management expenses overall. Chemical treatments increase the cost of producing textiles, and handling their recycling or disposal entails further expenses. While the environmental costs are not taken into consideration, this financial burden is frequently transferred to customers. A multifaceted strategy is needed to address these effects, one that includes the creation of eco-friendly finishing agents, sustainable dyes, and cutting-edge recycling technologies. Textile production can lessen its environmental impact while maintaining resource availability over the long term by giving environmental sustainability top priority.

3.7.2 Approaches to Mitigating Impacts on Recycling

3.7.2.1 Development of Eco-Friendly Dyes and Finishes

Nowadays, dyes derived from natural plants and sustainable finishing treatments are gaining interest by researchers in reducing the environmental impact. Eco-friendly dyes are developed as an alternative to traditional toxic dyes and finishing agents.

These dyes and finishes are generally non-toxic in nature, biodegradable, and easier to decolourize during the recycling process. By adopting these greener chemicals, the industry can minimize contamination and improve the recyclability of textiles (35).

3.7.2.2 Advanced Sorting and Separation Technologies

Sorting and separation techniques play a vital role in selecting the recycling techniques. But still, labour-oriented sorting persists leading to many human errors and being time consuming. Hence, advancement in the sorting and separation technology offers a beneficial solution to the challenges faced during recycling dyed and finished textiles. Innovations in automated sorting systems, such as spectral imaging and sensor-based technologies, enable more accurate identification and separation of materials based on their dye and finish compositions. This process can sort textiles by various parameters such as colour, material type, and chemical treatments, enhancing the efficiency of recycling processes.

3.7.2.3 Standardized Labelling and Certification

Standardized labelling and certification schemes can greatly enhance the ability to track and manage textiles with dyes and finishes. Labels showing exactly what chemicals have been used on the fabric and so on enable recyclers to pick out their materials by such markings for much more efficient processing. Certification schemes for sustainable dye and finish use get manufacturers involved in eco-friendly practices while also enlightening consumers about environmentally responsible choices.

3.7.2.4 Industry Collaboration and Regulation

An essential factor is teamwork between the fabric making industry, recycling places, and rule-makers to fix the problems dyes and finishing helpers have on reusing items. Rules shared by a group of industries instructing what should be done and offering ways for good working practices in making and remaking textiles can push more helpful materials, and methods onto everyone. Laws that set marks for how to use and throw away dyes or finishing agents will help, along with offering reasons to act in environmentally friendly ways, supporting the idea of a circular economy in textiles.

Dyes and finishing agents result in problems for the reprocessing of textiles; in addition, they create very enormous sustainability impacts. Chemicals to that effect should work towards their development and use, further recycling technologies, and assistance across the sector to help return the industry to a circular economy with minimized waste and lowered environmental impacts, while maximizing resource efficiency throughout the lifecycle of textiles. This will require continuous innovation within the industry as well as support towards meeting impacts on recycling from dyes and finishing agents for long-term sustainability through change embraced not only by the textile industry but also for leading a new future in sustainability.

3.8 SUMMARY

Technologies for recycling are essential for reducing waste and advancing environmental sustainability. Good waste management and recycling become increasingly difficult as the world's population rises and urbanization picks up speed. However, there is hope that these problems can be resolved thanks to recent developments in recycling technologies. The high initial costs of establishing recycling facilities, inadequate infrastructure for gathering and classifying recyclables, and a lack of funds for studies into more effective recycling techniques are some of the largest obstacles. Increasing public participation and awareness are still major challenges to improving the efficacy of recycling programmes.

In spite of these challenges, innovative new technologies such as sophisticated waste-to-energy systems and chemical recycling are making significant strides in addressing the global waste crisis. By improving the recycling procedure and the quality of recycled materials, these innovations are lessening the negative effects of waste disposal on the environment. For example, chemical recycling offers a more sustainable option to traditional recycling by converting synthetic waste into useful raw materials. In the meantime, waste-to-energy systems are assisting in the production of clean energy from waste, which lessens greenhouse gas emissions and our reliance on fossil fuels. Additionally, strategies like circular economy models and extended producer responsibility (EPR) are promoting sustainable production and consumption by guaranteeing that products are made to be recyclable from the beginning.

Cooperation between communities, businesses, and governments is crucial to overcoming the obstacles recycling technologies face. Governments ought to invest in improved recycling infrastructure and enact laws that provide incentives for recycling. More environmentally friendly production techniques must be used, and industries must be accountable for the post-use handling of their goods. Communities also play a significant role by promoting recycling locally and raising awareness of its significance. Research and development spending increases will also promote innovation and lower technical obstacles. Initiatives to raise public awareness can also support the development of a sustainable culture and motivate individuals to dispose of waste responsibly.

Reducing the environmental impact of waste and promoting a circular economy require the adoption of cutting-edge recycling technologies. Waste will no longer be thrown away or burnt; instead, it will be recycled and reused in the future if we embrace innovation and collaborate. Governments, businesses, and communities working together can overcome recycling obstacles and build a more sustainable world for coming generations.

REFERENCES

1. Kaza S, Yao LC, Bhada-Tata P, Van Woerden F. *What a Waste 2.0: A Global Snapshot of Solid Waste Management to 2050* [Internet]. Washington, DC: World Bank; 2018 [cited 2024 Nov 25]. Available from: https://hdl.handle.net/10986/30317
2. United Nations Statistics Division—Environment Statistics Section. (2020) Environment Statistics. Waste: Composition of municipal waste [Internet]. 2021. Available from: https://www.gov.uk/government/publications/guidance-on-applying-the-waste-hierarchy.
3. Department for Environment, Food and Rural Affairs. Guidance on applying the waste hierarchy [Internet]. 2011. Available from: https://www.gov.uk/government/publications/guidance-on-applying-the-waste-hierarchy

4. RoadRunner. The environmental crisis caused by textile waste [Internet]. 2021. Available from: https://www.roadrunnerwm.com/blog/textile-waste-environmental-crisis
5. Chioma. The environmental impact of textile waste: 5 ways it impacts our planet [Internet]. 2023. Available from: https://faircado.com/mag/the-environmental-impact-of-textile-waste-5-ways-it-impacts-our-planet/
6. Khandaker S, Bashar MM, Islam A, Hossain MdT, Teo SH, Awual MdR. Sustainable energy generation from textile biowaste and its challenges: A comprehensive review. *Renewable and Sustainable Energy Reviews*. 2022 Apr;157:112051.
7. Wang T, Liu B, Xue Y, Wang W, Chen S. Effect of textile waste on incineration behavior of dyeing sludge: Combustion characteristics, gas emissions, kinetics. *Journal of Cleaner Production*. 2024 Jan;435:140619.
8. Vela IC, Maric J, Seemann M. Valorisation of textile waste via steam gasification in a fluidized bed reactor. In: *HERAKLION 2019—7th International Conference*. Chalmers University of Technology; 2019.
9. Wealth in Waste: *India's Potential to Bring Textile Waste Back into the Supply Chain*. India: Fashion for Good Sorting for Circularity; 2022 Jul, p. 1–104.
10. Dahlbo H, Aalto K, Eskelinen H, Salmenperä H. Increasing textile circulation—Consequences and requirements. *Sustainable Production and Consumption*. 2017 Jan;9:44–57.
11. Millward-Hopkins J, Purnell P, Baurley S. Scenarios for reducing the environmental impacts of the UK clothing economy. *Journal of Cleaner Production*. 2023 Sep; 420:138352.
12. Wojnowska-Baryła I, Bernat K, Zaborowska M, Kulikowska D. The growing problem of textile waste generation—the current state of textile waste management. *Energies*. 2024 Mar 22;17(7):1528.
13. Textile Exchange. Preferred fiber & materials market report. [Internet]. 2022. Available from: https://textileexchange.org/knowledge-center/reports/preferred-fiber-and-materials/
14. Ekström KM, Salomonson N. Reuse and recycling of clothing and textiles—A network approach. *Journal of Macromarketing*. 2014 Sep;34(3):383–99.
15. Martikkala A, Mayanti B, Helo P, Lobov A, Ituarte IF. Smart textile waste collection system—dynamic route optimization with IoT. *Journal of Environmental Management*. 2023 Jun;335:117548.
16. David W, Anja G, Tova A, Pirjo H. Textile-to-textile recycling. Ten Nordic brands that are leading the way. [Internet]. Nordic Council of Ministers, Copenhagen, Denmark; 2017. Available from: http://www.norden.org/nordpub
17. Rieter. The increasing importance of recycling in the staple: Fiber spinning process. Rieter, 2020.
18. Athanasopoulos P, Zabaniotou A. Post-consumer textile thermochemical recycling to fuels and biocarbon: a critical review. *Science of the Total Environment*. 2022 Aug; 834:155387.
19. Sina Ogundaini R, Monsurat OT. Textile waste pyrolysis: an innovative method for petrochemicals generation for sustainable economic, technological and environmental advancement. *Environmental Science and Pollution Research*. 2024 Nov 26; 31(57):65627–37.
20. Hussain T. Chemical recycling of cotton waste: different technologies [Internet]. Sustainable and Circular Textiles. 2023. Available from: https://thetextilethinktank.org/chemical-recycling-of-cotton-waste-different-technologies/
21. Christoff-Tempesta T, Epps TH. Ionic-liquid-mediated deconstruction of polymers for advanced recycling and upcycling. *ACS Macro Letters*. 2023 Aug 15;12(8):1058–70.

22. Mills-Senn P. Chemical recycling may offer a solution to textile waste [Internet]. Fabric Architecture. 2023. Available from: https://fabricarchitecturemag.com/2023/11/01/chemical-recycling-may-offer-a-solution-to-textile-waste/
23. Minor AJ, Goldhahn R, Rihko-Struckmann L, Sundmacher K. Chemical recycling processes of nylon 6 to caprolactam: review and techno-economic assessment. *Chemical Engineering Journal.* 2023 Oct;474:145333.
24. Alberti C, Figueira R, Hofmann M, Koschke S, Enthaler S. Chemical recycling of end-of-life polyamide 6 via ring closing depolymerization. *ChemistrySelect.* 2019 Nov 22;4(43):12638–42.
25. Xi G, Lu M, Sun C. Study on depolymerization of waste polyethylene terephthalate into monomer of bis(2-hydroxyethyl terephthalate). *Polymer Degradation and Stability.* 2005 Jan;87(1):117–20.
26. Palhano Zanela TM, Curti Muniz E, Policiano Almeida CA. Chemical recycling of poly(ethylene terephthalate) (PET) by alkaline hydrolysis and catalyzed glycolysis. *Orbital: The Electronic Journal of Chemistry.* 2018 May 26;10(3):226–33.
27. Maurya A, Bhattacharya A, Khare SK. Enzymatic remediation of polyethylene terephthalate (PET)–based polymers for effective management of plastic wastes: An overview. *Frontiers in Bioengineering and Biotechnology.* 2020 Nov 19;8:602325.
28. Yu S, Li Q, Zhang Y, Zhou H. New possibility for PET plastic recycling by a tailored hydrolytic enzyme. *Green Energy and Environment.* 2024 Feb;9(2):163–5.
29. Lefeuvre A, Garnier S, Jacquemin L, Pillain B, Sonnemann G. Anticipating in-use stocks of carbon fibre reinforced polymers and related waste generated by the wind power sector until 2050. *Resources, Conservation & Recycling.* 2019 Feb;141:30–9.
30. Pakdel E, Kashi S, Varley R, Wang X. Recent progress in recycling carbon fibre reinforced composites and dry carbon fibre wastes. *Resources, Conservation & Recycling.* 2021 Mar;166:105340.
31. Howarth J, Mareddy SSR, Mativenga PT. Energy intensity and environmental analysis of mechanical recycling of carbon fibre composite. *Journal of Cleaner Production.* 2014 Oct;81:46–50.
32. Szatkowski P, Twaróg R. Thermal recycling process of carbon fibers from composite scrap—characterization of pyrolysis conditions and determination of the quality of recovered fibers. *Fibers.* 2024 Aug 21;12(8):68.
33. Pickering SJ, Yip H, Kennerley JR, Kelly R, Rudd CD. The recycling of carbon fibre composites using a fluidised bed process. In: Centre for Composite Materials Engineering, University of Newcastle, UK, editor.*FRC 2000–Composites for the Millennium* [Internet]. Cambridge, UK: Woodhead Publishing; 2000 [cited 2024 Dec 25]. p. 565–72. Available from: https://linkinghub.elsevier.com/retrieve/pii/B9781855735507500724
34. Dang W, Kubouchi M, Yamamoto S, Sembokuya H, Tsuda K. An approach to chemical recycling of epoxy resin cured with amine using nitric acid. *Polymer.* 2002 May;43(10):2953–8.
35. Pizzicato B, Pacifico S, Cayuela D, Mijas G, Riba-Moliner M. Advancements in sustainable natural dyes for textile applications: A review. *Molecules.* 2023 Aug 8;28(16):5954.

4 Digitalization of Tracking and Separation Technology for Recycling

4.1 INTRODUCTION

The textile industry is a multi-national web of various stakeholders like fibre, yarn, and fabric producers, dyers and finishers, garment manufacturers, accessory providers, and retailers. The vast supply chain frequently cuts across several nations, which renders the recycling and management of textile materials a difficult task. In such a scenario, digital technologies, specifically targeted at identification, tracking, tracing, and separation, have become central to the industry as a whole. Such tools are crucial to the process of attaining a circular economy (CE) in textiles, wherein the materials are recycled, reused, and reclaimed into the production loop.

Digital technologies have an important role during the transition period by making recycling processes more efficient and effective. For instance, computer-aided sorting units based on artificial intelligence (AI) and machine learning algorithms assist in distinguishing and sorting textile waste materials based on their composition, colour, and quality. Such precise sorting assures that recycled materials from textile waste have the desired standards for reuse in new products. Radio frequency identification (RFID) and near field communication (NFC) tags, along with blockchain, enable real-time tracking and traceability of textiles across their lifecycle, improving recycling efficiency (1). Digital Product Passports (DPPs) further support circularity by providing detailed product information to aid reuse, disassembly, and material recovery.

The textile sector's shift towards digital technologies is driving the growth of circular economies built on digital solutions. These systems connect manufacturers, consumers, and recyclers through shared platforms, making it easier to exchange information and resources.

4.2 COMPLEXITIES OF WASTE RECYCLING

Identifying fibre composition in a product containing mono fibre or fibre blends is critical for CE, specifically recycling. Traditionally, fibres have been identified using microscopy, chemical dissolution, or thermal analysis such as burning tests and calorimetry. However, these techniques are slow, require high labour and skilled technical expertise, and often cannot cope well with fibre blends, coatings, dyes, or degraded/aged textiles. Laboratory-based tests can take several hours per sample and are costly, and facilities may not be available in a sorting facility, making them impractical for high-throughput waste streams. In the textile recycling sector, due to

DOI: 10.1201/9781003503835-4

the lack of fast, cost-effective, and reliable tests, identification and segregation are still challenging and often carried out manually, relying on the experience of workers, which is highly unreliable.

4.3 IDENTIFICATION OF FIBRES FOR RECYCLING

Identification of fibres is a crucial step in textile recycling, as accurate composition analysis determines the recycling route and quality of recovered materials. Advances in non-destructive spectroscopic techniques, coupled with AI technology, are increasingly demonstrating high accuracy, throughput, and reliability.

A technology developed for textile waste sorting is Raman spectroscopy combined with machine learning, analyzing the molecular vibration of the polymer and grouping textile waste into six classes based on fibre composition. The system can sort at about one piece per second passing through a conveyor belt, whereas manual sorting takes more time for one piece (2).

The most promising technology for rapid identification is near-infrared spectroscopy (NIR), which uses light in the near-infrared region. In comparison, Fourier Transform Infrared (FTIR) spectroscopy uses mid-infrared (MIR) light and provides information only about surface properties, whereas NIR analyzes bulk samples. NIR (e.g., Fibersort system) is faster, easier to handle, requires no sample preparation, and is better suited for industrial environments, while FTIR is still mostly confined to laboratory studies. Handheld NIR spectrometers used for qualitative identification (e.g., distinguishing cotton vs polyester) also show high reproducibility.

NIR-based identification technology is advancing rapidly, especially when combined with machine learning for NIR wavelength recognition for fibre classification. These systems can segregate textiles by fibre type, and when combined with colour sensors, also by colour. By training systems on known NIR wavelength datasets, AI is now used to recognize fibre blends of varying proportions with high accuracy.

However, detection may be hindered by coatings, finishes, fabric thickness, and fabric age. For example, one pilot study using NIR in an automated recognition line achieved only 73% accuracy over ~250 textile fabric pieces when coatings, finishes, blends, thin fabrics, prints/dyes, etc. were present (3).

A study classifying pure fibres (cotton, linen, wool, silk, polyester, polyamide, viscose) achieved 100% accuracy for mono-fibres using NIR individually. Moreover, when blends are involved (e.g., cotton/polyester, polyester/viscose, or including elastane), spectral overlap makes it hard to distinguish fibre content, especially when minor fibre fractions are small or when surface treatments obscure signature peaks. An AI-driven data fusion model (combining both NIR + FTIR spectra) captures complementary chemical information from both spectra. Machine learning algorithms were trained on this fused data to recognize subtle spectral differences. This improves accuracy and reduces classification errors to as low as ~5% in worst cases (about two misclassifications out of 45 samples) (4).

NIR signals are processed using multivariate statistical tools such as Principal Component Analysis (PCA), Soft Independent Modelling of Class Analogy (SIMCA), and Linear Discriminant Analysis (LDA). These methods enable accurate identification of fibres such as polyamide, viscose, polyester, silk, linen, and cotton. Using NIR spectroscopy with multivariate calibration (e.g., PLS regression, etc.), models have been developed that can estimate the cotton content in cotton-polyester blends (5).

Commercial instruments using NIR are now being developed to combine identification with robotic sorting. Examples include Fibersort and SIPTex. Adoption of these rapid sorting technologies still remains limited due to cost. High capital and operational costs of hyperspectral cameras, spectrometers, robotics, and AI infrastructure, along with challenges of maintenance, calibration, and sensor drift, hinder scalability. There is scope to improve the systems for identification of mixed and blended fabrics of wide fibre types and with different fabric finishes (6, 7).

4.4 DIGITAL PRODUCT PASSPORT

The textile supply chain is a multifaceted network with numerous tiers and countries and has a buyer-driven production approach. There is a need for comprehensive information for consumers to make informed purchasing decisions. Information is also needed for resale and recycling. DPP offers detailed information about each product throughout its lifecycle to promote circular practice, traceability, and transparency. The European Commission defines the **Digital Product Passport** as "a combination of an identifier, the granularity of which can vary throughout the lifecycle (from a batch to a single product), and data characterizing the product, processes, and stakeholders, collected and used by all stakeholders involved in the circularity process."

DPPs enable tracking and tracing, including after-sale use and end-of-life management. They typically contain information on:

- Product description and composition
- Supply chain and transport documentation
- Environmental, social, and animal welfare impacts
- Circularity and health impacts
- Brand information
- Quantity and cost data
- After-sale tracking and tracing
- Customer feedback

It enables reliable, secure mapping of products from fibre to consumer. It guarantees authenticity and accountability across the value chain, influencing design, sample development, sourcing, manufacturing, distribution, retail, returns, and recycling.

Key benefits include:

- Greater supply chain transparency and consumer confidence
- Authenticity verification and trade dispute resolution
- Prevent counterfeiting
- Support for regulators and certifying bodies
- Monitor social and environmental responsibility (e.g., child labour, sanctions, fair wages)
- Provide repair, care, and reuse guidance, reducing wear and tear
- Support return logistics through smart packaging and tracking

4.4.1 Types of Textile Digital Technologies and Platforms

Growth of digitization is seen mostly in online resale platforms, where product composition, history, and sustainability information guide consumer decisions. These platforms—including Depop, Vinted, eBay, ASOS Marketplace, Facebook Marketplace, and Etsy—play an important role in recommerce and fashion rental markets. These platforms are growing and supporting resale of clothing. *Seamless*, the world's first clothing stewardship programme in Australia, reported that in 2024, 60 million items were reused in Australia, a 10% increase from 2023.

The success of digitization depends on broad technology adoption, cost-effectiveness, accuracy, reliability, and consumer awareness. Encouragingly, advances in machine learning, AI, and mobile platforms are driving progress in clothing supply chains. Blockchain-based platforms, such as Textile Genesis and others, are increasingly integrated into DPPs to trace fibres from origin to consumer. Bext360 uses satellite imagery, sensors, and blockchain to track cotton production from farm to consumer. Reverse Resource is a digital brokerage platform providing real-time data on textile waste flows, enabling multi-stakeholder engagement and textile-to-textile recycling. While promising, these technologies face challenges in scalability, standardization, and governance.

4.4.2 Tracing and Tracking Technologies

Transparent and traceable supply chains are critical for CE. These technologies fall into three main categories:

4.4.2.1 Forensic Tracers

The use of forensic analytical methods helps to understand the inherent biochemical fibre composition. This is important for microfibre identification from oceans to determine the type of textiles from which the particles have come. The methods include elemental analysis, DNA analysis, and microbiome profiling. DNA-based tracing is accurate in differentiating fibres. An example includes DNA-based tracing that distinguishes cashmere from wool, which are otherwise difficult to identify (8).

4.4.2.2 Additive Tracers

Additive tracers are those added to fibres and/or garments during production via nano-mists, inks, pigments, or digital serialization. Detection instruments can be handheld or in-line UV light detectors or apps on smartphones. Molecular identifiers are added to cotton fibre during ginning. Some digital watermarks and nano invisible inks are added to garment labels, but those embedded into the fibre are durable during use and repeated washing to assist post-consumer traceability for recycling. Other examples of additives are artificial DNA, optical fingerprinting, and rare earth fluorescent nanoparticles.

In the cotton sector, DNA markers are attached to cotton fibres at the farm level, allowing for tracking and verification at various stages of the supply chain by a DNA test. They can provide a high level of accuracy and security, as they are unique and difficult to tamper with. However, the implementation challenges include costly, complex, and time-consuming verification processes. DNA technology can be used in other fibres like wool or even synthetics.

Data management technology goes hand in hand with tracing and tracking technologies. For example, in **Fibre Trace** technology, specialized luminescent pigments are added to the cotton fibres, creating a unique signature that can be scanned and tracked throughout the supply chain. Each time a scan occurs, FibreTrace® generates location data, date, time, and auditor information, which are securely stored with FibreTrace's blockchain technology. Similarly, the **Haelixa** platform is used to collect, analyze, and share the data generated by the DNA marker. Challenges include cost, complexity, and time-consuming verification.

4.4.3 Tags

UV- or IR-legible fibres or quick response (QR) codes and radio frequency identification (RFID) tags are embedded into yarns so they can withstand use and washing, even if labels are removed. The thread containing RFID technology informs about the product composition for easy recycling. RFID is long-lasting, suitable for rapid reading of bulk items, and not easily damaged. Unlike RFID, barcodes and QR codes are degradable but can be made waterproof and durable to be suitable for clothing products (9).

An RFID reader can be placed in a wardrobe, helping to register the number of times items are taken out and used, as well as reminding the wearer to wear "idle" clothes. This is useful for sharing and rental platforms as well as recycling (9, 10).

4.5 SEPARATION OF FIBRES FROM OTHER ITEMS IN CLOTHING FOR RECYCLING

Efficient separation of fibres from non-fibre components in clothing is a critical step in textile recycling, directly influencing the quality and viability of recycled materials. Clothing often contains mixed fibres such as cotton-polyester blends, along with added components like zippers, buttons, dyes, and finishing chemicals. Traditional manual sorting methods cut off zippers, buttons, and labels before shredding the garment. This process is labour-intensive but ensures better fibre quality for recycling. Industrial shredders can partly separate fabric from hard items like zippers or rivets. After shredding, magnets (for metal items) and density-based separation systems help remove non-fibre parts. In some advanced processes (e.g., hydrothermal or solvent-based), fibres are dissolved or depolymerized, leaving behind non-fibre residues like zippers and buttons. These residues can then be filtered out more easily.

Modern facilities are experimenting with robotic arms and AI-driven vision systems to identify and remove trims like buttons, rivets, or zippers before fibre recycling. AI-driven sorting systems help to increase recycling rates based on their application and material types. These systems not only increase efficiency but also improve material purity by accurately identifying and segregating different fibre types and contaminants.

4.6 SEPARATION OF BLENDS FOR RECYCLING

Separation of blended fabrics, such as cotton-polyester (cotton-PET) or cotton-elastane, remains one of the most challenging aspects of textile recycling. These blends are prevalent in the fashion industry due to their desirable properties like durability and stretch, but they complicate recycling processes because the individual fibres are tightly interwoven and chemically distinct.

Traditional mechanical recycling methods struggle to effectively separate these fibres, often resulting in downcycling into lower-value products like insulation or rags. Chemical recycling technologies, such as H&M's Green Machine, aim to address this issue through a hydrothermal process that combines heat, water, and biodegradable chemicals to efficiently separate cotton from polyester. This process allows for the recovery of both materials: polyester can be spun into new yarn, and cotton is processed into cellulose powder. Although the economic viability and scalability of such technologies are often concerns due to high investment and infrastructure needs, companies like BlockTexx have successfully adapted their recycling technology on a commercial scale.

Although recycling blended textiles remains a major challenge, ongoing research and new technologies are steadily opening the door to more effective solutions. The integration of advanced sorting technologies and chemical recycling methods holds promise for improving the sustainability of the textile industry. However, achieving widespread adoption will require continued investment, collaboration, and development of scalable solutions.

4.7 SUMMARY

Recycling textile waste is complex because of the wide variety of fibres, blends, and chemical additives such as dyes, flame retardants, and finishes. Unlike materials like plastic bottles, textile fibres are harder to recycle as their additives hinder fibre recovery and demand advanced sorting and identification methods. Traditional identification techniques, including microscopy and chemical digestion, are often slow, destructive, and poorly suited for modern blended fabrics.

Recent advances in digital technology are addressing these challenges. Tools such as near-infrared (NIR) spectroscopy, hyperspectral imaging, and AI-driven vision systems make fibre identification faster, non-destructive, and more precise. They allow real-time separation of materials, reducing contamination and improving the quality of recycled output. Alongside this, DPPs are emerging as powerful tools for circularity. By storing detailed data about a product's materials, production, and recycling practices, DPPs give recyclers, producers, and consumers the information they need for efficient material recovery and reuse.

Another critical step is the separation of non-fibre components like buttons, zippers, and elastics. Digitalized systems using robotics, sensors, and AI can handle this automatically, cutting labour needs and ensuring cleaner fibre streams. Blended fabrics such as cotton-polyester remain particularly challenging, but digitally assisted methods such as optical sensors or laser technology are helping preserve fibre integrity and improve downstream applications.

Despite these advances, several barriers limit large-scale adoption. Concerns about data privacy, lack of platform standardization, and difficulties integrating digital systems into existing recycling infrastructure need to be addressed. Progress will depend on innovation, supportive policies, and collaboration across the textile value chain. By adopting digital solutions for tracking, identification, and separation, the textile industry can enhance recycling efficiency, conserve resources, and reduce environmental impact, moving significantly closer to a sustainable CE.

REFERENCES

1. Nayak R, Singh A, Padhye R, Wang L. RFID in textile and clothing manufacturing: technology and challenges. *Fashion and Textiles*. 2015 Jun 2;2(1):9.
2. Tsai PF, Yuan SM. Using infrared Raman spectroscopy with machine learning and deep learning as an automatic textile-sorting technology for waste textiles. *Sensors*. 2024 Dec 25;25(1):57.
3. Cura K, Rintala N, Kamppuri T, Saarimäki E, Heikkilä P. Textile recognition and sorting for recycling at an automated line using near infrared spectroscopy. *Recycling*. 2021 Feb 8;6(1):11.
4. Riba JR, Cantero R, Puig R. Classification of textile samples using data fusion combining near- and mid-infrared spectral information. *Polymers*. 2022 Jul 29;14(15):3073.
5. Paz ML, Sousa C. Discrimination and quantification of cotton and polyester textile samples using near-infrared and mid-infrared spectroscopies. *Molecules*. 2024 Aug 2;29(15):3667.
6. Riba JR, Cantero R, Canals T, Puig R. Circular economy of post-consumer textile waste: Classification through infrared spectroscopy. *Journal of Cleaner Production*. 2020 Nov;272:123011.
7. Liu Z, Li W, Wei Z. Qualitative classification of waste textiles based on near infrared spectroscopy and the convolutional network. *Textile Research Journal*. 2020 May;90(9–10):1057–66.
8. Zhang X, Wu X, Yang H, Zheng H, Zhou Y. Identification of cashmere and wool by DNA barcode. *Journal of Natural Fibers*. 2023 Apr 24;20(1):2175100.
9. Agrawal TK, Koehl L, Campagne C. A secured tag for implementation of traceability in textile and clothing supply chain. *The International Journal of Advanced Manufacturing Technology*. 2018 Dec;99(9–12):2563–77.
10. Perlacia AS, Duml V, Saebi T. Collaborative consumption: live fashion, don't own it: developing new business models for the fashion industry. *Beta*. 2017 May 12;31(1):6–24.

5 Recent Advances in Recycling Technologies: Case Studies

5.1 INTRODUCTION

Recycling is essential for advancing sustainability and lessening the negative effects of waste on the environment. Significant advancements in recycling technologies have been made in tandem with technological advancements to increase the efficacy and efficiency of the recycling process.

5.2 SORTING TECHNOLOGIES

One of the most challenging phases of textile waste recycling is sorting the waste based on its composition. Generally, manual sorting was carried out before recycling, but it is time consuming and less accurate. These kinds of issues have led to advancements in sorting techniques such as tray sorters, separation robots, and garment-on-hanger (GOH) sorting systems. These innovations have simplified the process with minimal error.

5.2.1 Tray Sorters

Tray sorters are one of the automated textile waste-sorting techniques where the textile waste is separated and collected in compartments based on different parameters such as colour, type, and quality. Firstly, the textile waste from various sectors is combined and loaded onto a conveyor belt. Later, the materials are scanned using sensors or cameras to identify the categories of waste, which are then directed to the appropriate compartments. Near-infrared (NIR) spectroscopy is used for textile identification such as blend proportion of two different fibres and determining the types of fibres by their chemical composition (1). For example, cotton clothing might be placed on one tray, polyester on another, and mixed fibres on a third. The tray sorter arranges the items in the respective compartment according to the analysis provided by the sensors. STADLER, a top recycling plant supplier, in association with TOMRA, developed a pilot textile waste-sorting plant in Sweden in 2017. As part of the SIPTex project, this led to a larger plant in 2020—the first fully automated system that sorts mixed textile waste for effective recycling and reuse. This sorter is located in Malmo, which sorts both pre-consumer and post-consumer waste, which has other metallic parts like zipper, buttons, etc. The sorting process is significantly quicker and more accurate when using automated sorting as opposed to manual

DOI: 10.1201/9781003503835-5

methods. Especially effective are tray sorters for large-scale operations where accuracy and high throughput are essential. They are helpful tools in the recycling industry because of their capacity to handle a broad range of textiles including fabrics for clothing and homes.

5.2.2 Separation Robots

Introduction of robots has upgraded the technology used for textile sorting. Their system uses sensors, AI, and machine learning to organize fabric waste into categories. The fact that robots designed for separation handle and control each single item means they are ahead of tray sorters. Equipped with various sensors, the robots inspect the item's presentation, how high it is, and information about its material. Robots can become more accurate at their task by using AI after being told what to improve. They have the ability to manage updates in textile classification and products. A major benefit of separation robots is their ability to deal with challenging sorting activities. Examples include the ability to separate items that have several layers or use complicated materials, which ordinary methods fail to tackle. Also, such robots are able to prepare items for the next task, recycle them, or handle delicate tasks such as labelling or trimming (2).

5.2.3 Garment-on-Hanger Sortation Systems

GOH sortation systems are used to arrange garments hung on hangers. GOH sortation tools are designed to gather together clothes left on hangers. Both textile recycling and managing inventory at retail and fashion businesses are part of the current strategy. GOH uses hangers so the garments are hung and then carried further along the company's conveyor system. Classifying by colour, composition, type, and size of each clothing item is done using sensors and RFID tags. Only now does the clothing undergo final delivery to places such as sorting bins or processing zones. With GOH systems, it is easy to set aside small items that can either be sold or used once more. The method avoids unnecessary touch and damage since the clothes remain on hangers. As these methods are highly efficient, a lot of clothing can be sorted rapidly.

5.2.4 Supervised Learning

The labelled data used to train supervised learning algorithms pairs the input data with the appropriate output. These algorithms gain knowledge from this training data in order to forecast or decide on fresh unobserved data. Supervised learning can be used in textile sorting to recognize various fabric types, colours, and patterns. These are examples of popular supervised learning algorithms.

- **Support vector machines (SVM):** SVMs are able to categorize textiles according to features like texture and colour patterns that are taken from photos.

- **Random forests:** To increase classification accuracy, this ensemble learning technique employs several decision trees. Textiles can be sorted using it according to a number of criteria including material type, condition, and category for recycling or intended reuse.
- **Neural networks:** For image recognition tasks, deep learning models—in particular convolutional neural networks or CNNs—are very successful. Through the analysis of visual features, CNNs are able to accurately identify and classify textile materials.

5.2.5 Unsupervised Learning

Without being specifically told what to look for, unsupervised learning algorithms use unlabelled data to find patterns and structures. These algorithms can be used to find hidden patterns in textile waste like recurring damage types or common fabric blends. Important methods for unsupervised learning include:

- **Clustering algorithms:** To help with the effective classification of mixed textile waste, algorithms such as k-means and hierarchical clustering can combine similar textile items according to their characteristics.
- By reducing the dimensionality of the data, **Principal Component Analysis (PCA)** highlights the key characteristics that set various textile types apart. Focusing on important characteristics can help to expedite the sorting process.

5.2.6 Reinforcement Learning

Algorithms that use reinforcement learning pick up new skills through interaction with their surroundings and feedback in the form of rewards or penalties. In dynamic sorting systems where the best sorting strategy may evolve over time, this method can be especially helpful. For example:

- **Q-learning:** By determining which actions (e.g., guiding an item to a particular bin) provide the best results (e.g., precision and swiftness).
- **Deep reinforcement learning:** By fusing deep learning and reinforcement learning, these models are able to manage intricate decision-making tasks in real-time like modifying sorting plans in response to the flow of textile waste.

The "Transform Textile Waste into Feedstock" (TTWiF) initiative, led by TEXAID, endeavours to assess and elevate the cutting-edge technologies for sorting textile waste across Europe. This initiative aims to formulate a blueprint for advanced textile sorting facilities, pinpointing the most effective technologies, systems, and methodologies to boost efficiency and integration. The research concentrates on three primary domains:

- **(Semi-)Automated sorting for reuse:** Sorting mixed clothing, shoes, and accessories for reuse is a time-consuming procedure that is typically done by hand and requires several steps to categorize items by type and quality, sometimes into more than 300 categories. Accuracy and efficiency can be greatly increased by automation and digital improvements like sensor-based detection systems, automated handling, and transportation technologies. Features like product type, condition, size, brand, style, material, colour, textile construction, and trims should all be recognized by these systems. Additionally, clothing without information tags or with digital product passports (DPP) should be scanned by the technology.

- **Automated sorting for recycling qualities:** At the moment, manual sorting based on composition labels or material feel when available does not guarantee accurate material composition, particularly for clothing with multiple layers or mixed fibres. Clothing must be sorted by fibre composition by automated systems in order to facilitate efficient fibre-to-fibre recycling. In order to precisely identify all parts and trims, clothing must be scanned from the inside out.
- **Automated pre-processing for recycling:** Pre-processing is necessary to eliminate contaminants such as trims and other components for fibre-to-fibre recycling to be effective. Cutting tools that are semi-automated or manual can be used to remove trims. Complex clothing with many layers and trims, however, can be difficult to work with and may end up being downcycled or burned. For precise fibre-to-fibre recycling, a two-step sorting procedure is required. To start, a spiked lattice conveyor is used to feed mixed textiles, breaking up clumps and distributing them uniformly. Conveyor belt sensors identify the composition of the textiles, after which air pressure nozzles classify the materials and compress them into bales for recycling or storage. Cutting down textiles into smaller pieces prior to reassessment by sensors improves the material purity. Here, random cutting is carried out followed by compressed air sorting and sensor-based detection. This stage removes contaminants like metals and plastics to guarantee precise feedstock requirements for recycling.

5.3 MECHANICAL RECYCLING

New technology in mechanical recycling has improved the effectiveness, quality, and scalability of recycling textile waste. These improvements come from upgraded shredding and extrusion which help transform post-consumer and post-industrial textile waste into useful raw materials for production.

5.3.1 Improved Shredding Techniques

5.3.1.1 Automated Sorting and Shredding Integration

Nowadays, shredders are frequently teamed up with equipment designed to sort materials by themselves. Before processing, these machines are capable of separating the textile material from other types. Sorting your material in advance stops it from polluting your shredder and makes the outcome of your shredding far better.

5.3.1.2 High-Efficiency Shredders

Modern shredders, with their flexibility, allow customers to recycle both everyday clothing and fabric mixtures. The blades on these shredders are strong enough to transform fabric into equal and tiny particles, making it easier for recycling.

5.3.1.3 Multi-Stage Shredding

In order to create finer and more uniform fibre outputs, multiple shredding processes have been developed. The process turns the textile into large pieces and then into segments that are finer once or several times. Waste fibres can become high-quality recycled materials if they are first cleaned, sized, and separated; a multi-stage approach improves the efficiency of these steps.

5.3.1.4 Dust and Fibre Recovery Systems

Shredders are also responsible for recycling dust and fibre. They separate out the small particles produced, helping to maintain recycling. Since they cut down on airborne fibres, these machines increase worker comfort and recover more resources.

5.3.2 Advanced Extrusion Processes

5.3.2.1 Twin-Screw Extrusion

The use of new twin-screw extrusion methods makes combining shredded textile fibres and compound materials simpler. A twin-screw extruder is especially useful in recycling since it can produce different fibre products. Because of the uniform outflow from the extruders, produced material can be used for innovative textile products.

5.3.2.2 Reactive Extrusion

Reactive extrusion uses chemicals inside the extruder to reform the features of the textile fibres. The properties of the recycled substance can be enhanced by choosing different fibres and introducing useful chemicals to it. Reactive extrusion is a way to recycle polyester and nylon synthetic fibres.

5.3.2.3 Continuous Filament Extrusion

For making continuous, high-quality fibres out of used textiles, processes for extrusion have been carefully adjusted. In order to use recycled fibres to make yarn and non-woven materials, the process combines melting these scraps and flowing them out through specialized spinnerets. Recycled textile materials are still useful for many different products once continuous filament production is applied.

5.3.2.4 Energy-Efficient Extrusion Systems

There has been a recent emphasis on making extrusion processes more energy efficient. Energy usage is kept to a minimum by adding advanced heating systems, improved screw shapes, and upgraded control electronics in modern extruders.

5.3.2.5 Enhanced Pelletizing Techniques

When recycled fibres are extruded, many times they are first pelletized for better management and processing. Enhanced pelletizing methods such as strand and underwater pelletizing mean the pellets are more consistent and meet the quality standards set for use in many textile production lines.

5.4 CHEMICAL RECYCLING AND BLEND SEPARATION

Textile waste can now be converted into useful raw materials thanks to recent advancements in chemical recycling technologies, especially solvent-based procedures and depolymerization techniques. These developments make it possible to recycle synthetic and blended fibres that are difficult to process with mechanical recycling techniques, which is crucial for handling the complex and diverse nature of textile materials.

5.4.1 Solvent-Based Processes

5.4.1.1 Selective Dissolution

In solvent-based recycling, fibres are separated from all types of textiles. Studies in exclusive dissolving techniques for polyester or nylon to protect other fibres from harm are a new trend. Precipitated pure polymers can be gathered from the solution, used once more to produce new textiles, and recovered again by applying this method (3). Environmentally safe types of solvents are used (3).

5.4.1.2 Environmentally Friendly Solvents

People are paying a lot of attention to making environmentally safe solvents. Solvents are now created so they can be recycled, reused, are less hazardous, and easier to dispose of. Excellent efficient polymer dissolution with minimal environmental issues can be provided by ionic liquids and supercritical fluids (4, 5).

5.4.1.3 Closed-Loop Systems

Due to new developments in solvent recycling, many processes have systems that recycle and use the same solvents during repeated dissolution. Disposing of fewer solvents helps the environment and also brings down the company's overall spending.

5.4.2 Depolymerization Techniques

5.4.2.1 Glycolysis

Polyester polymers are recycled by glycolysis, which separates them into their building blocks with the help of ethylene glycol. The latest research on glycolysis has focused on choosing the best reaction and catalyst conditions to get better results (6). By cleaving small molecules from the polyester, the purity of the monomers goes up, and they can be turned into new types of polyester fibres (6).

5.4.2.2 Hydrolysis

Hydrolysis breaks down polymers when water and catalysts are used. Researchers have studied how temperature, pressure, and the amount of catalyst influence hydrolysis in efforts to make the degradation of polyesters and nylons more efficient. Using specific enzymes to catalyze the process seems to offer a good way to produce high-quality results while using less heat or pressure (7).

5.4.2.3 Methanolysis

In methanolysis, polyester is broken down with methanol into dimethyl terephthalate (DMT) and ethylene glycol (EG). Both the use of continuous reactors and better catalysts are examples of new methods in methanolysis. For example, polyethylene terephthalate (PET) treated using glycolysis and then methanolysis produces DMT. Thus, this technique can now help industry complete its work both effectively and inexpensively (8).

5.4.2.4 Ammonolysis

Researchers use ammonolysis more often now to separate the bonds in polymers by exposing them to ammonia. Recent research is being done to test if ammonolysis will turn nylon into caprolactam, which can then be processed into new nylon fibres. In order for this technique to work well in industry, more effective catalysts and optimized processes are necessary (9).

5.5 DEPOLYMERIZATION AND RECYCLING OF SYNTHETIC POLYMERS

Depolymerization is a promising way to change synthetics into monomers, allowing waste plastics to be used in creating more plastic. Due to microwave heating and enzymatic reactions, recycling is now an option for a larger group of plastics, like PET and nylon.

5.5.1 Microwave-Assisted Depolymerization

In microwave-assisted depolymerization, polymers are heated by microwaves to divide them into their basic monomer units. It has a number of benefits when compared to the standard thermal depolymerization method. When compared to traditional methods, microwave-assisted depolymerization is efficient and environmentally friendly. Using heating from all sides tends to give uniform heat distribution and decreases depolymerization time more than traditional techniques. As a result, companies can use less electricity, are more eco-friendly, and need fewer people to run the plant, which cuts costs. In addition, microwaves help set precise reaction conditions, including temperature and pressure, so that more of the monomers are recovered with improved purity. Because it can handle many polymers, including PET and nylon, this technique is ideal for recycling various plastic waste packing boxes.

5.5.2 Enzymatic Degradation

To break down polymers in enzymatic degradation, specific enzymes are used to unlock the bonds between the large and small parts. Because this method chooses specific molecules and takes place under mild conditions, it brings several important advantages. By using enzymes to break plastics, we have a safer choice for recycling than traditional chemical methods. An important advantage is that enzymes work solely on their target molecules. As a consequence, they can remove polymers of different types without affecting other materials, making the monomers very pure and suitable for use in fresh plastics. Most enzymatic methods excel well at high temperatures and pressures, distinct from classic methods that use these pressures and temperatures, which strongly affect the environment. That is why heat pumps are sustainable and cost less to operate. Additionally, the chemicals in enzymes break down over time and are safe for the environment, so you will not have to add extra chemicals. Because of recent progress in biotechnology, scientists can now target a greater variety of synthetic polymers by finding and changing enzymes. Because of this, experts are turning to the use of enzymes to help recycle many types of plastic. Since it is efficient, safe for nature, and changeable, this system may address the worldwide problem of plastic waste.

5.6 OTHER RECYCLING TECHNOLOGIES USED IN THE TEXTILE INDUSTRY

Recent advancements in recycling technologies for the textile industry have significantly improved the efficiency and sustainability of textile waste management. Key innovations include water purification systems, dye recovery processes, and chemical recycling methods, each contributing to a more circular and eco-friendly textile industry.

5.6.1 Water Purification Systems

5.6.1.1 Closed-Loop Water Recycling

New water purification systems are meant to recycle water during the textile manufacturing process. With these systems, wastewater is processed, removing harmful particles, so the water can be reused in the dyeing and finishing parts of the process. As a result, water is used less and less safe chemical is disposed of into the environment.

5.6.1.2 Advanced Filtration Technologies

Membrane filtration, ultrafiltration, and nanofiltration are now widely adopted to separate particles, dyes, and contaminants from wastewater. With these technologies, recycled water can be purified sufficiently to be reused by consumers.

5.6.1.3 Zero Liquid Discharge Systems

Zero Liquid Discharge (ZLD) works to keep wastewater out of the environment by recovering both water and solid waste used during textile processing. Because of this system, less water is consumed and waste from the textile process is minimized, so any remaining leftovers are either reused or treated with care.

5.6.2 Dye Recovery Processes

5.6.2.1 Solvent-Based Dye Recovery

Solvent-based methods are applied to recover and clean dyes from wastewater. This method involves dissolving dyes in a suitable solvent and then removing them from water by methods such as distillation or filtration with membranes. Instead of throwing away dyes, recovered ones can be put back into textile manufacturing, saving new dyes and reducing the amount of dye waste.

5.6.2.2 Enzyme-Based Dye Degradation

Generally, using enzymes for breaking down dyes into particles is less harmful. It serves well to decolourize contaminated water before treating it for use again. Eco-friendly enzymes can break down dyes and are available for a wide range of dyes.

5.6.2.3 Electrodialysis and Electrocoagulation

Electrochemical techniques are applied to collect dyes and important substances from wastewater. With electrodialysis, electrical energy is used to split ionic compounds, whereas electrocoagulation forms clumps that capture and take out impurities. Through either method, the dye recovery rate goes up and water quality improves.

5.6.3 Chemical Recycling Methods

5.6.3.1 Solvent-Based Recycling

Synthetic fibres are dissolved in solvents in solvent-based chemical recycling so that the original monomers can be recovered. With the help of processes including solvent extraction and selective dissolution, textile materials with a high impurity level are prepared for use in new garments.

5.6.3.2 Depolymerization Techniques

Depolymerization converts the polymers into monomers by breaking their chemical bonds. Such techniques as glycolysis, hydrolysis, and methanolysis are applied to recycle PET and nylon textiles. Lately, researchers have honed both catalysts and environmental conditions to boost efficiency and the ability to apply technology at a large scale.

5.6.3.3 Pyrolysis and Gasification

In pyrolysis, textile waste is heated in an oxygen-free environment to produce valuable gases, oils, and char. Gasification uses high doses of heat and some oxygen to convert textile waste into syngas (hydrogen and carbon monoxide). Both support the change of textile waste into resources that people can use.

5.6.3.4 Enzymatic Recycling

Enzymes are used to break up textiles into short polymer chains or useful items. Using enzymes, you can recycle challenging materials more effectively and sustainably than with standard processes that make use of chemicals. Developments in enzyme engineering have enabled more types of fibres to be handled using enzymes.

5.7 CASE STUDY 1: PET BOTTLES RECYCLING INTO TEXTILE MATERIALS

Shree Renga Polyester Pvt Ltd. (SRPPL) has emerged as a frontrunner in PET bottle recycling and sustainable textiles in Karur, Tamil Nadu. With a focus on transforming post-consumer PET bottle waste into high-quality recycled polyester staple fibres, yarns, and garments, SRPPL addresses environmental challenges while delivering value-added products. This case study delves into the company's operations, achievements, innovative products, and vision for a sustainable future.

5.7.1 Company Overview

SRPPL's operations are rooted in a strong commitment to sustainability and innovation. In 2008, the company launched its PET bottle recycling line, which was expanded by 1.5 times in 2010. To further its eco-friendly initiatives, SRPPL introduced dope-dyed technology in 2011, enabling waterless colouring and significantly reducing water and air pollution. The company produces dope-dyed recycled fibres (Renga Fibres), dope-dyed coloured polyester spun yarns (Renga Yarns), fabrics, and garments. In 2021, SRPPL founded Ecoline, India's first sustainable fashion brand.

The company operates six facilities: a crushing unit, a polyester staple fibres dope-dyeing unit, a fabric and garment manufacturing unit, a research and development centre, and a greenfield project for finished goods. With a collection network spanning Tamil Nadu, Kerala, Pondicherry, Karnataka, and Andhra Pradesh, SRPPL recycles over 1.5 million PET bottles daily. It has also achieved significant product development milestones, including the creation of over 200 colours and the development of specialized fibres such as flame-retardant, antimicrobial, tri-lobal, and hollow fibres.

- 15 lakh PET bottles are used every day for producing 25 tons of recycled fibres.
- The waste caps are used for battery case manufacturing and road laying, whereas waste wrappers are used by cement factories.
- Water used in the process of cleaning the PET bottles is recycled through reverse osmosis process with ZLD process. Nearly 29 billion litres of water have been saved to date.
- The fibre follows dope-dyeing technology that does not use a single drop of water, making it the most eco-friendly way of colouring polyester.
- Fine denier recycled polyester fibres are sold to textile spinning mills to produce mélange yarn, OE yarns, and blended yarns.
- Coarser denier fibre is supplied to automotive companies like Hyundai, Kia, Mahindra and Mahindra, Tata Motors, and Toyota for making floor carpets and roof laminates.
- Coarser denier fibres are also sold to companies making abrasives like Scotch-Brite.

5.7.2 Recycling Process

1. **Bottle collection and sorting:** Bottles are collected from the entire Tamil Nadu, parts of Kerala, Pondicherry, Karnataka, and Andhra Pradesh. Also, PET bottles are collected from diverse locations, including challenging terrains like the Siachen Glacier, in collaboration with the Indian Army. Bottles are sorted and segregated for non-PET and other contamination.
2. **Cleaning and shredding:** Bottles pass through a series of Trommel, label remover, and conveyors to remove labels, caps, and other residual contamination. Finally, bottles are crushed into flakes of size 12–14 mm. The sizes are critical for the further melting process in spinning. Dirty flakes are washed in a continuous line using caustic soda at 95°C and rinsed. In this process, paper bits, floatable contamination, and glue are removed. Washed and rinsed flakes are dried and then conveyed to the melt spinning section.
3. **Drying process:** PET flakes are dried to a moisture level of 50 ppm in continuous drum dryers and column dryers as PET depolymerizes in the presence of moisture. Dryers are maintained at 190°C to drive out tough moisture by adsorption method using molecular sieves. Dried flakes are then fed to the extruders for melt spinning.
4. **Fibre production:** Flakes are melted along with requisite colour master batch in an extruder and the melt is filtered and spun into fibres or filament yarns through fine spinnerets as shown in Figure 5.1.
5. Spinneret holes from 0.25 to 0.5 mm decide the final denier. Depending upon the applications, the denier of the filament can be fixed. Spun filaments are stretched in a series of rollers to achieve final denier and strength. In case of staple fibre production, the filament in the form of tow is cut into

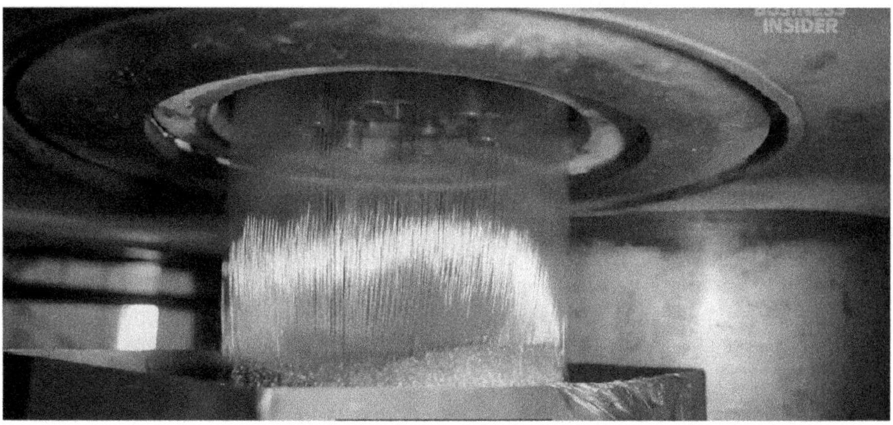

FIGURE 5.1 Fibres extruded through spinneret holes.

desired length as per the order. In the case of filaments, it is stretched four times in a draw twisting machine to arrive at the final denier and strength. Normally, 1.2 denier to 15 denier fibres are produced for textiles, automotive, and industrial woven/non-woven applications. Cut length of staple fibre will vary from 32 to 76 mm depending on the yarn requirement. In the case of filaments, 75 denier to 300 denier are produced for the same kind of applications.

6. **Yarn production:** Dope-dyed staple fibres either alone or in blend with cotton (50:50, 60:40, 70:30) are spun in ring frame or open end machine to produce yarn of different counts. 20s and 30s are preferred for garment applications, whereas 10s are used in home textile applications.
7. **Fabric and clothing manufacture:** Yarn in the form of spun or filament is either woven or knitted to make fabrics for garmenting. Yarn can be 100% or blended with cotton/wool/nylon/acrylics at various ratios for the desired applications. Fabrics are then finally stitched into garments for various end uses.

5.7.3 Recycled Polyester Yarn (Renga Yarn)

- The recycled polyester fibre produced from PET bottles is used as a primary component in producing recycled yarn. The fibres are subjected to ring spinning technique to convert them into ring spun yarn.
- More than 50 plus colour shades and 10 plus count variants of spun yarn were produced.
- Following specifications are maintained in spun yarn:
  - Twist per inch (TPI): 16–20
  - Hairiness less than 7.5
  - Twist multiplier: 4.2
  - Thin, thick, and neps are under control
  - Count strength product (CSP): 4,000–4,500
- Recycled polyester yarns produced in different counts ranging from 4, 6, 10, 16, 20, 24, 30, 36, to 40s are sold to home textile applications.
- Recycled yarn is supplied to companies to make ticking fabrics for mattresses.

5.7.4 Ecoline Clothing

- SRPPL manufactures garments and sells them under a brand name called "Ecoline Clothing" (see Figure 5.2). Ecoline is truly a futuristic fashion brand that has its deepest focus etched on producing sustainable clothing solutions.
- All the garments are made from knitted fabrics in different structures such as Double tuck pique, Single Jersey, loop knit, and interlocks. The GSM

FIGURE 5.2 Sewing lines of Ecoline Clothing.

(grams per square metre) normally ranges from 160 to 250 depending on the style of garments. Typically, 190 GSM is used to make a Polo T-shirt.

- Garments for men, women, and children are sold in various colours and styles. Sports goods are produced and treated with wicking properties, antimicrobial, anti-odour, anti-static, UV protection, and soft feel properties.
- Ecoline range of products is of export quality, and they are much loved in American and European nations.
- Waist coat is the most popular dress that was decorated by our honourable Prime Minister and other VIPs in India and abroad.

5.7.5 Life Cycle Assessment (LCA)

- No water is used for dyeing the fibres
- 1 r-PET T-shirt saves 3.7 kg of CO_2 equivalent
- 1 waist coat saves 8 kg of CO_2 equivalent
- 6 PET bottles recycled to 1 r-PET T-shirt
- 28 PET bottles recycled to 1 jacket
- 45 PET bottles recycled to 1 blazer
- 2.9 billion litres of water saved
- 2.4 lakh tons of CO_2 saved
- Renewable energy resources such as solar and windmills were used.

5.7.6 SRPPL Operation Flow Chart

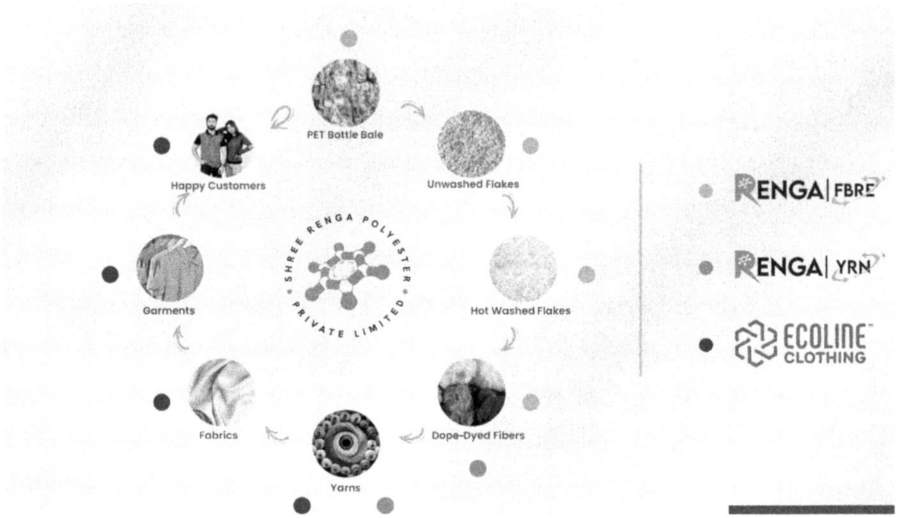

FIGURE 5.3 Process flow of Sree Renga Polyester Pvt. Ltd.

5.7.7 Flushbric

- Flushbric is an innovative eco-friendly brick and paver block made from effluent sludge waste of PET bottle recycling plant and the fly ash from the steam boiler.
- Sludge, boiler ash, aggregates, and cement are mixed in a proprietary ratio and moulded in a continuous machine to produce paver blocks. These blocks are cured for 14 days to achieve required compressive strength.
- Flushbric is certified in a NABL lab for compressive strength.

5.7.8 Achievements

- Collaborated with Clean Bhutan to produce traditional Bhutanese dresses (Kho and Kira) from PET waste.
- Explored partnerships with countries like Sri Lanka, Qatar, Maldives, and Nepal to establish recycling plants.
- Certified by global quality standards, including:
  - Global Recycled Standards for traceability
  - Oekotex for colour stability
  - ISO 9000 and 14000 for quality and environmental impact
  - ZED certification for zero defects and zero effects
  - OPB (Ocean Bound Plastics) for ocean-collected bottles

5.7.9 Research and Development Projects

The company is actively engaged in a range of innovative research and development projects aimed at advancing sustainability and functionality. It has developed bi-component filament lines tailored for sports garments and baby products, showcasing its commitment to high-performance textiles. By recycling bottle caps, the company has created durable products such as vegetable crates, battery cases, and large 3D-printed components. It has also ventured into producing non-woven fabrics for nursery grow bags and elite multi-use bags crafted from vegan leather derived from sugarcane bagasse. Collaborations with academic institutions have led to the development of stronger paver blocks and acoustic boards. Additionally, the company is exploring projects to produce biodegradable polyester and blended fabrics incorporating natural fibres like Tencel, bamboo, and Modal, further underscoring its dedication to innovative and eco-friendly solutions.

5.7.10 Startup Initiatives

SRPPL has mentored and invested in startups focused on sustainability, including:

- Carbon Masters, specializing in kitchen waste recycling to produce biogas and organic manure.
- Startups supporting farmers with natural products like mangoes, free-range eggs, and fresh milk.

5.7.11 Challenges and Solutions

5.7.11.1 Challenges

- Ensuring consistent raw material supply.
- Competing with cost-effective conventional polyester producers.
- Raising consumer awareness about recycled textiles.

5.7.11.2 Solutions

- Established robust collection networks for PET bottles.
- Conducted awareness campaigns to promote sustainable practices.
- Invested in advanced technologies to optimize production costs and quality.

Shree Renga Polyester Pvt Ltd. exemplifies how sustainable practices can transform waste management and textile production. By leveraging innovation, community engagement, and environmental stewardship, SRPPL serves as a model for achieving economic success while contributing to a greener planet. SRPPL envisions scaling its recycling capacity, diversifying product offerings, and expanding globally. The company is committed to continuous innovation in sustainable textiles and aims to strengthen its role as a leader in the circular economy.

5.8 SUMMARY

Using recent recycling technologies on textile waste, together with new chemical, mechanical, and water purification methods, has brought significant progress towards an eco-friendly textile industry. These advancements can treat textile waste in a better, more scalable, and greener way every year. Bringing these technologies together is necessary to get the most from them and build a complete system for recycling. Gathering parts from recycled clothes with mechanics, followed by chemical processing, increases the chances of generating premium recycled fibres. A combination of leading sorting methods and recycling processes using either solvents or enzymes is an effective approach to managing mixed textile waste.

For technologies to keep developing, studies and experiments must not stop. Growth in enzyme engineering, increased solvent reclamation, and using different recycling methods together will lead to more efficient, scalable, and sustainable processes. Better teamwork among textile manufacturers, suppliers of technology, scientists, and lawmakers will lead to faster use and implementation of these advanced techniques. Combining knowledge and assets can improve solutions and encourage the development of a circular economy. Taking circular economy steps such as designing recyclable products, organizing take-back initiatives, and boosting closed-loop methods will make room for recycling technologies and see reclaimed materials returned into the supply chain. Suitable rules and policies strongly influence the growth of innovation. Encouraging recycling and giving money for research help drive new and improved practices in recycling in the textile industry.

If people learn about the value of recycling textiles and are invited to join recycling activities, recycling programmes are much more likely to achieve success. Informing consumers about sustainability and showing them how to play a part can increase interest in recycled textiles and support the industry's efforts. Overall, modern technologies in recycling textile waste are helping make the textile industry more sustainable. By emphasizing integration, innovation, and cooperation and supporting policies that promote circular economy ideas, the industry can keep advancing in solving the environmental issues related to textile waste. As more people use these technologies, they will shape the way textile waste is handled and make the world more sustainable.

REFERENCES

1. Cura K, Rintala N, Kamppuri T, Saarimäki E, Heikkilä P. Textile recognition and sorting for recycling at an automated line using near infrared spectroscopy. *Recycling*. 2021 Feb 8;6(1):11.
2. HKRITA. H&M Foundation. 2024. AI-powered Smart Garment Sorting System automating textile recycling. Available from: https://hmfoundation.com/project/ai-powered-smart-garment-sorting-system-automating-textile-recycling/#:~:text=7-,AI%2Dpowered%20Smart%20Garment%20Sorting%20System%20automating%20textile%20recycling,How%20it%20works
3. CreaCycle GmbH. Der CreaSolv Prozess. [Internet]. 2023. Available from: https://www.creacycle.de/de/der-prozess.html

4. Wang Q, Lu X, Zhou X, Zhu M, He H, Zhang X. 1-Allyl-3-methylimidazolium halometallate ionic liquids as efficient catalysts for the glycolysis of poly(ethylene terephthalate). *Journal of Applied Polymer Science.* 2013 Sep 15;129(6):3574–81.
5. Christoff-Tempesta T, Epps TH. Ionic-liquid-mediated deconstruction of polymers for advanced recycling and upcycling. *ACS Macro Letters.* 2023 Aug 15;12(8):1058–70.
6. Sheel A, Pant D. Chemical depolymerization of PET bottles via glycolysis. In: Sabu T, Ajay Vasudeo R, Krishnan K, Abitha VK., Martin George T, editors. *Recycling of Polyethylene Terephthalate Bottles* [Internet]. Oxford, UK: Elsevier; 2019 [cited 2025 May 14]. p. 61–84. Available from: https://linkinghub.elsevier.com/retrieve/pii/B9780128113615000043
7. Papadopoulou A, Hecht K, Buller R. Enzymatic PET degradation. *CHIMIA.* 2019 Sep 18;73(9):743.
8. Xu WH, Chen L, Zhang S, Du RC, Liu X, Xu S, et al. New insights into urethane alcoholysis enable chemical full recycling of blended fabric waste. *Green Chemistry.* 2023;25(1):245–55.
9. Suhaimi NAS, Muhamad F, Abd Razak NA, Zeimaran E. Recycling of polyethylene terephthalate wastes: A review of technologies, routes, and applications. *Polymer Engineering & Science.* 2022 Aug;62(8):2355–75.

6 Textile Waste: Value Addition and Applications

6.1 INTRODUCTION

The fashion and textile industries produce massive amounts of textile waste annually and take little action to control it. The environment is under a lot of stress because of the waste, and managing it requires a lot of work and money. This difficulty can be overcome by pursuing cutting-edge tactics like upcycling, which is gaining popularity. Upcycling is the process of turning unwanted textile materials into functional goods. In addition to upcycling, these techniques reuse, recycle, and repurpose waste to create new business opportunities while also adding value and protecting the environment.

Recycling specifically attempts to lessen the environmental impact by conserving natural resources and turning used textiles into fibres for new yarns and fabrics. Within the context of reuse, used materials are transformed into useful products like bags, quilts, and home décor. However, repurposing changes the material's original purpose, such as turning used T-shirts into cushion covers or cleaning rags.

As sustainability gains international attention, there is a growing trend towards creating high-value goods from textile waste. Globally, it is estimated that between 25% and 30% of textile waste is recycled and used to make geotextiles, packaging, insulation, and automobile interiors. However, recycling rates vary by region. In contrast to the United States, which recycles 15% of its textile waste, the European Union recycles roughly 25%. Certain policies and incentives have been implemented in some nations to promote greater recycling rates. For instance, 30% of India's textile waste is recycled to create insulation mats and blankets. Japan, which has the highest rate at 50%, credits its success to stringent waste management regulations and efficient sorting systems.

The handcrafted applications of upcycled textile waste are truly outstanding in the fashion and interior design industries. These materials are not only constructed for eco-friendly garments and accessories, but also serve as active porous materials in industrial settings—for instance, soundproofing and insulation alongside home furnishings. Adding environmental value through textile waste gives companies an opportunity to address the growing market need for greener products. Recycled textiles have now technologically advanced to include uses such as agricultural textiles, oil absorbents, packaging, filtration systems, construction materials, and thermal or acoustic insulation (1). The need for innovation in sustainability alongside their

DOI: 10.1201/9781003503835-6

post-consumer and industrial waste allows the fashion and textile industries to decrease the sustainability footprint economically and environmentally. This innovative way of waste management not only helps reduce the reliance on virgin resources but also curbs waste bound for landfills. This results in decreased energy expenditure and greenhouse gas emissions, which significantly lowers the carbon footprint. In addition, this approach preserves natural resources.

Social entrepreneurship and community engagement contribute significantly towards the sustainable use of textile waste. Employment opportunities and new business models can be encouraged through guided projects involving sorting, classifying, and recycling textiles. Income generation through the enhanced participation of local artisans in the manufacture of geotextiles is veritable, and so is the advancement of environmental responsibility.

Substantial infrastructure and advanced technology are required for sorting and processing textile waste, which is a labour-oriented activity as well as costlier. The quality of the textile waste materials makes it challenging to produce fibres equivalent to virgin fibres. Despite these challenges, the reuse of textile waste for value-added products is a viable option. Industries can benefit by reducing the environmental impact, by using resources effectively, and by meeting the increasing demand for eco-friendly products.

6.1.1 Different Categories of Recycled Fibres Quality

Generally, recycled fibres are classified into three quality ranges based on their quantity:

- **High-quality recycled fibres (20–25%):** Fibres that retain good tensile strength, flexibility, and durability. This type of fibre is more suitable for producing high-end products like clothing, upholstery, and durable packaging. Derived from well-preserved pre-consumer waste, such as manufacturing off-cuts and unused fabrics.
- **Medium-quality recycled fibres (40–50%):** Moderate strength and durability; fibres may have some degradation. Used for mid-range products such as non-woven fabrics, industrial wipes, and insulation. Mostly sourced from post-consumer waste, such as discarded garments that are still in good condition.
- **Low-quality recycled fibres (25–35%):** Weaker, less durable fibres, often mixed with other materials to improve strength. Suitable for low-end applications like stuffing, cushioning, and industrial mats. Often derived from heavily used post-consumer textiles, which may be worn out or damaged.

The process of recycling textile waste, especially post-consumer textiles, results in shorter and weaker fibres than the original material. Consequently, blending with virgin fibres or using chemical treatments to enhance fibre quality is sometimes necessary to make recycled products that meet the required standards for durability and performance.

6.1.2 Quality of Recycled Fibres from Textile Waste Raw Materials

Common textile materials such as cotton, wool, silk, and polyester exhibit different properties before and after recycling. Understanding these characteristics helps determine their suitability for various applications.

6.1.2.1 Cotton Waste

- **Before recycling:** Cotton fibres typically range from 20 to 40 mm (short-staple cotton) to 50 mm or more (long-staple cotton). High tensile strength, making it suitable for durable fabrics. Cotton fibres are fine, with a diameter of around 12–20 μm. Used for clothing, bedsheets, and high-end textiles.
- **After recycling:** Recycled cotton fibres are shorter, often reduced to around 10–20 mm, which affects the overall strength and smoothness of the yarn. The process of recycling degrades the fibres, reducing tensile strength. As a result, the recycled cotton is generally less durable. The fibres may become coarser and less uniform due to the recycling process. Recycled cotton is often blended with virgin fibres for making products like insulation, stuffing, industrial wipes, and low-grade textiles.

6.1.2.2 Wool Waste

- **Before recycling:** Wool fibres vary widely, ranging from 25 to 150 mm, depending on the type of wool (e.g., Merino vs. coarse wool). Wool fibres are relatively strong, elastic, and have natural crimp, which enhances their insulation properties. Wool fibres range from 17 to 25 μm (fine wool) to 35 μm or more (coarse wool). Used in high-quality garments, upholstery, and carpets.
- **After recycling:** Recycled wool fibres are shorter, generally ranging between 15 and 50 mm, which affects their ability to be spun into fine yarns. Wool loses some of its natural elasticity and strength during the recycling process, resulting in weaker fibres. The fibres may become coarser, and blending with other fibres may be necessary to improve texture. Recycled wool is typically used for making blankets, industrial felt, and low-grade apparel, often blended with other fibres.

6.1.2.3 Silk Waste

- **Before recycling:** Silk fibres are continuous and can measure up to 600 m in length, which is why silk has such a smooth texture. Silk has high tensile strength and natural lustre, making it ideal for luxurious fabrics. Extremely fine, with a diameter of about 10 μm. Used for high-end clothing, scarves, ties, and accessories.
- **After recycling:** Recycled silk fibres are much shorter, often reduced to less than 50 mm, which results in a loss of smoothness. The recycled silk loses some of its strength, flexibility, and lustre, making it less suitable for premium applications. The fineness is reduced, and the texture becomes rougher. Recycled silk is often used for creating blended fabrics, decorative items, and some upholstery.

6.1.2.4 Polyester Waste

- **Before recycling:** Polyester fibres can be produced in both long filament and staple fibre form, typically 30–150 mm for staple fibres. Strong, durable, and resistant to stretching and shrinking. Can be fine to medium, depending on the manufacturing process, usually ranging around 10–25 μm. Widely used in clothing, home furnishings, and industrial textiles.
- **After recycling:** Recycled polyester fibres are usually shorter, averaging around 25–50 mm for staple fibres. The recycling process can cause a slight reduction in strength, but advances in technology have allowed recycled polyester (r-PET) to retain much of its original durability. The fineness may remain relatively consistent if the recycling process is controlled well. However, slight coarseness can occur. Recycled polyester is used in clothing (especially sportswear), home furnishings, packaging, and industrial applications. It is often blended with virgin polyester to improve quality.

The recycling of textile waste is an essential step towards achieving sustainability. However, the quality of fibres can change significantly during the recycling process. For cotton, wool, and silk, the fibres tend to become shorter, less strong, and coarser, affecting their application. On the other hand, recycled polyester has shown great potential, retaining much of its original qualities, especially when processed with advanced techniques. By understanding the properties of recycled fibres, industries can determine their suitability for various applications, from creating new fabrics and packaging materials to industrial products like insulation, automotive interiors, and geotextiles. As recycling technologies advance, the gap between the quality of virgin and recycled fibres is expected to narrow, leading to broader adoption of recycled materials across multiple industries.

6.2 THERMAL INSULATION

6.2.1 Modes of Heat Transfer

Thermal insulation is the method of reduction of heat transfer between objects or environments of varying temperature that are in thermal contact or within the heat transfer zone. Heat transfer is the movement of thermal energy from one object or environment to another when there is a temperature difference. This occurs as a result of conduction, convection, and radiation, which are the three primary mechanisms of heat transfer. **Conductive heat transfer** occurs when the transfer of heat takes place through direct contact. It occurs primarily in solids. This occurs when one part of a material is heated, the particles in that region gain energy and vibrate more rapidly, and these energized particles collide with the neighbouring particles, transferring energy (heat) through the material. **Convective heat transfer** refers to the transfer of heat caused by the movement of fluids (liquids and gases) from one area to another. Convection occurs when a fluid is heated, causing its molecules to gain energy and become less dense; as a result, warm fluids rise, allowing cooler dense fluids to move in and replace the warm fluid's space. This occurs in a loop, creating a pattern known as a convection current. This process transfers heat throughout the fluid, warming cooler areas as the fluid circulates. **Radiant heat transfer** takes place through electromagnetic waves. Unlike conduction and convection, radiation does not need a medium (solids and fluids). This occurs as a result of all objects emitting thermal radiation according to their temperature. Hotter objects or environments emit more radiation than cooler ones. When these waves encounter another object or environment, they can be absorbed, reflected, or transmitted based on their properties.

Heat energy: "One kilocalorie (1,000 calories or 1 kcal) is the amount of heat energy required to raise the temperature of one kilogram (kg) of water by one degree Celsius (°C)." The SI unit for energy is Joule (J). **Thermal conductivity:** It is the property that explains the ability of a material to transfer heat from one object or environment to another. This thermal conductivity, *K*-value, is expressed in units of W/m-K (watts per metre Kelvin). Thermal conductivity value depends on the material's average temperature and moisture content in it. Generally, materials with lower thermal conductivity indicate a more effective insulating property, meaning they prevent heat from passing through efficiently. Thermal conductivity values allow for a quantitative comparison between the effectiveness of different thermal insulation materials.

$$K = \frac{Q \times d}{A \times \Delta T}$$

where, *K*—thermal conductivity; *Q*—amount of heat transferred; *d*—distance between two isothermal planes; *A*—area of surface; ΔT—difference in temperature.

Thermal resistivity: Thermal resistivity is the property that describes the ability of the material to resist the transfer of heat from one object or environment to another. Thermal resistivity is the reciprocal of *K*-value (m · K/W). **Thermal resistance (*R*-value)** is a measure of the resistance offered by a material to heat flow. It is a function of material thermal conductivity, thickness, and density. The higher the *R*-value, the lower the heat loss and the more efficient the insulation of the material. It is expressed in m^2-K/W.

6.2.2 Thermal Insulation Application in Apparels

Thermal insulation is a key property in textiles, reducing the transfer of heat between surfaces and providing warmth and comfort to the wearer. Textile waste is developed into a lofty structure by non-woven technology (thermal bonded) to have thermal insulation properties, and a range of new value-added applications can be achieved. One approach is using recycled fibres and fabrics to create new thermal insulation materials. Another method involves the development of insulating materials by incorporating waste wool, down feathers, or synthetic fibres into the textile waste, enhancing its thermal properties. These added materials create loft within the textile structure, trapping air and reducing heat transfer, which provides effective insulation. In the home textiles sector, these materials can be used to create cosy blankets, comforters, and pillows that offer warmth and comfort. In the apparel industry, value-added textile waste can be utilized to make insulated jackets, coats, and other cold-weather garments.

6.2.3 Thermal Insulation Application in Buildings

Thermal insulation materials find use in various industrial applications, such as providing insulation for buildings, thermal barriers for industrial equipment, and protective covers for machinery. Thermal insulation plays a vital role in maintaining an apt environment within the building for human comfort. Reports say the majority of heat losses in an unprotected building are distributed with approximately 40% through walls, 47% through windows and glass, and 13% through the roof and ground (1). The role of insulation is highlighted by the fact that buildings and the construction sector collectively account for about 32% of total energy consumption and contribute to about one-third of total carbon dioxide emissions. Most emissions are due to the combustion of fossil fuels for heating, cooling, lighting, and powering appliances and electrical equipment. Studies reveal that good insulation in a building can save up to 65% of total energy expenditure (2). After installing thermal insulation panels to the building, findings showed a 55% reduction in heating energy consumption during winter and an 18% decrease in air conditioning usage in summer (3). The thermal insulation market, valued at USD 71.7 million in 2023, is anticipated to reach USD 96 million by 2028, with a projected compound annual growth rate (CAGR) of 6%.

6.2.4 Working Mechanism of Thermal Insulation Materials

Thermal insulation is essential for enhancing energy efficiency in buildings by reducing heat gain and loss through the building envelope. A basic level of thermal insulation is necessary to shield building components from thermal stress and moisture-related deterioration. During winter, the primary purpose of thermal insulation is to conserve energy, which in turn reduces the need for heating and helps protect the environment. This aim has to be considered in new buildings as well as in renovating the building stock. Strategies to reach this aim include the use of building materials with low thermal conductivity values so as to minimize the heat interaction between internal and external surroundings. Thermal insulation material provides a region in which heat transfer is either eliminated or minimized, creating a thermal barrier. The material does this by trapping tiny pockets of still air. This air provides thermal insulation rather than the material itself, as still air is a good insulator of heat with thermal conductivity of about 0.016 W/m K. Additionally, creating small cells (closed cell structure) within thermal insulation material aids reduction of heat transfer through radiation. When the material has these small cells, the paths that heat would normally take are broken up, resulting in absorption and scattering instead of passing through the material.

Thermal insulation in buildings is a crucial element in energy efficiency, as it reduces the transfer of heat between the inside and outside environment. This aids in maintaining a stable and comfortable indoor climate, thereby lowering the reliance on heating during winter and cooling in summer. The mechanism of thermal insulation is based on controlling heat transfer through conduction, convection, and radiation as shown in Figure 6.1. **Conduction** is the transfer of heat through a solid material. In buildings, heat can pass through walls, roofs, and floors by conduction. Thermal insulation reduces this by using materials with low thermal conductivity,

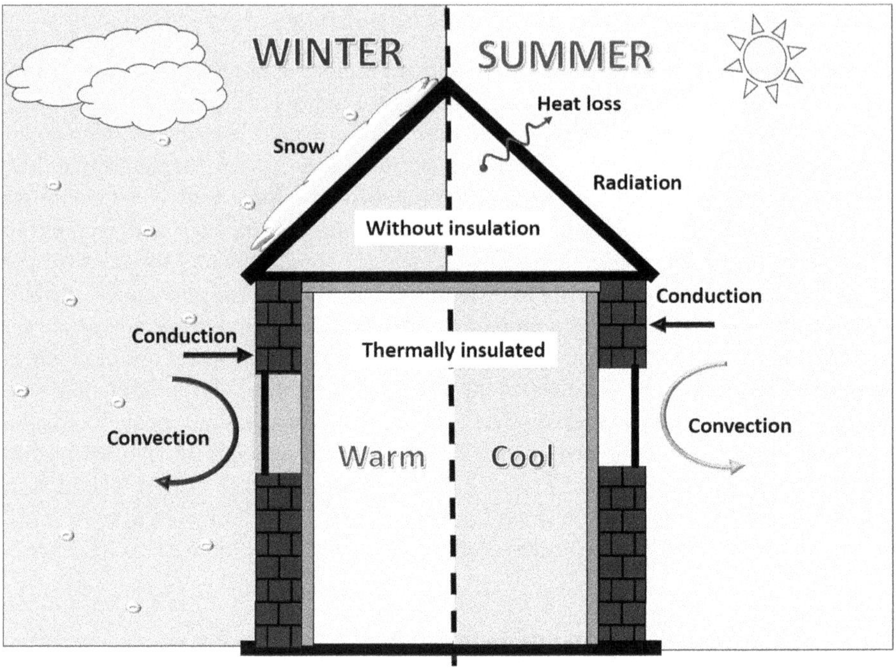

FIGURE 6.1 Mechanism of thermal insulation in building application.

meaning they do not easily allow heat to pass through. **Convection** involves the transfer of heat through fluids (liquids and gases). In buildings, air can carry heat away from surfaces, causing heat loss. Insulation minimizes this by trapping air within small pockets, preventing the free movement of air and thus reducing heat loss. **Radiant** heat transfer occurs when heat is emitted from a warm surface and absorbed by a cooler one. This can be mitigated using reflective insulation materials, such as foil, which reflects radiant heat back towards its source.

6.2.5 Sustainable Thermal Insulation Panel

Beyond manufacturing new products, there is an opportunity to incorporate value-added textile waste with thermal insulation properties into sustainable building practices. Insulation produced from recycled textile materials can enhance a building's energy efficiency by lowering heating and cooling expenses and minimizing the environmental footprint of construction activities. This not only supports sustainable construction practices but also promotes the use of recycled materials in mainstream applications. Non-woven materials based on waste and recycled fibres have immense potential to be used as sustainable insulation materials (4), with properties comparable to those of conventional insulation materials such as rock wool, glass wool, and mineral wool, which are based on non-renewable inorganic resources, while foam-based insulation materials are based on petroleum-based raw material sources, which are fast-depleting resources. Non-woven structures with adequate layering arrangements can act as a viable eco-friendly alternative to commercial insulators.

6.2.5.1 Case Study 1: Thermal Insulation Non-woven Using Recycled PET and Comber Noil

A promising research area is the use of recycled PET (polyethylene terephthalate) fibres for thermal insulation, which are mostly obtained from plastic bottles that have been used after consumption (5). Both chemical and mechanical methods are used to recycle these fibres, but even though chemical processes yield higher quality, mechanical recycling is preferred because it is less expensive. One economical way to obtain recycled PET (r-PET) fibres with qualities like high strength and extensibility is through mechanical recycling, which entails melting the plastic and then re-extruding it into yarns. To enhance the properties of r-PET fibres for specific applications, such as thermal insulation, surface modification is often employed. In this study, recycled PET fibres are treated with sodium hydroxide to modify their surface properties. The process revealed a roughened fibre surface due to the presence of excess oligomers after alkali treatment, enhancing the fibres' suitability for subsequent processing into non-woven materials.

The development of needle-punched non-woven materials is carried out by two approaches: one with 100% recycled PET fibres and another by blending PET fibres with comber noil (CN), a waste by-product from cotton yarn production. The developed non-wovens exhibited a thermal conductivity ranging between 0.017 and 0.020 W/m K, indicating their potential for thermal insulation applications (see Figure 6.2).

It is observed that the alkali-treated fibres showed slightly lower thermal conductivity values due to their enhanced surface characteristics. However, the introduction of CN at varying blend percentages adversely affected thermal insulation, attributed

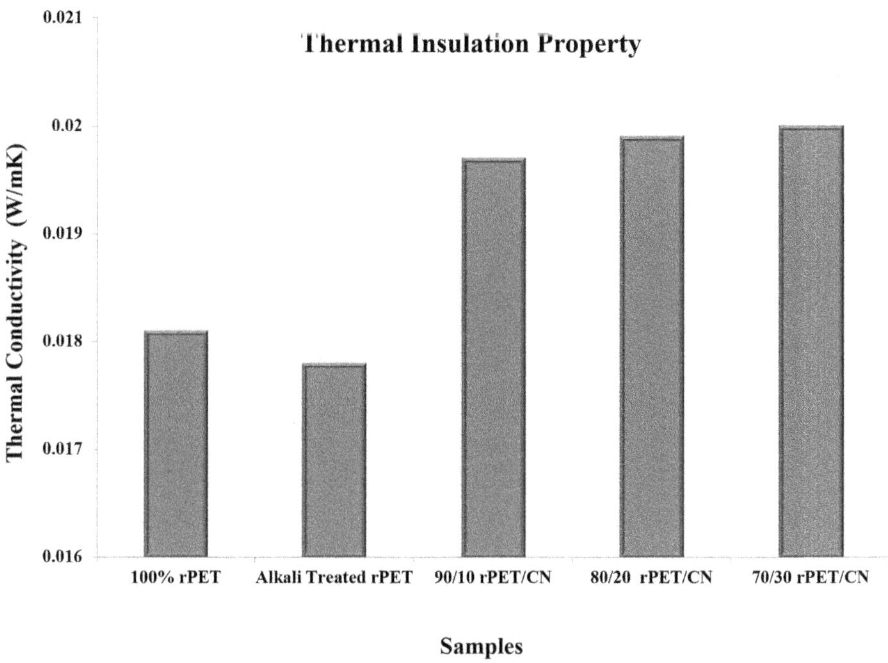

FIGURE 6.2 Thermal insulation properties of recycled PET/comber noil non-wovens.

to increased moisture content. Higher moisture content in fibres leads to greater heat sorption, thereby reducing the thermal resistance of the non-woven materials. Future research could explore optimizing blend ratios of recycled PET fibres with CN to mitigate the impact on thermal insulation while enhancing other desirable properties. The study underscored several environmental benefits associated with the use of recycled PET fibres, which reduces the cost of raw materials.

6.2.5.2 Case Study 2: Thermal Insulation Non-woven Using Banana Fibre and Recycled Polyester

This study investigates the development of needle-punched and chemically bonded non-wovens using banana fibre and recycled polyester (r-PET) fibre blends (6). The r-PET fibres are derived from recycled plastic bottles and blended with banana fibres to enhance mechanical and thermal properties. The research focuses on thermal insulation and biodegradation qualities that are affected by various blend ratios and bonding methods. Utilizing the needle-punching and chemical bonding techniques, the non-wovens are created using blends of banana fibre and r-PET in three distinct ratios: 90/10, 70/30, and 50/50.

All the developed non-wovens showed excellent insulation properties, with thermal resistance values ranging from 0.299 to 0.248 $m^2 \cdot K/W$ (see Figure 6.3). Interestingly, the 90/10 non-woven blend exhibited better thermal insulation performance than other blend ratios across both methods. Furthermore, chemically bonded non-wovens exhibited a 6.69% greater thermal resistance compared to needle-punched non-wovens, emphasizing the significant impact of bonding techniques on insulation performance. This may be attributed to the use of adhesives to interlock

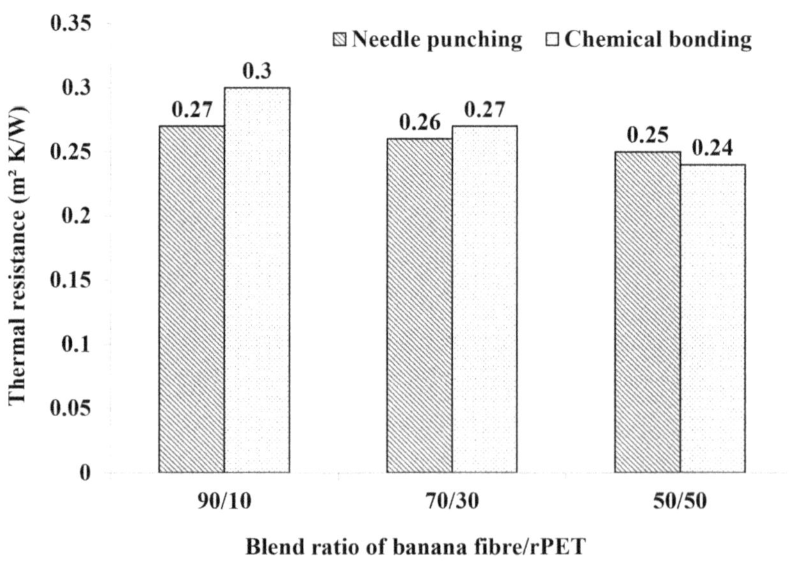

FIGURE 6.3 Thermal resistance of non-woven developed from different blend ratio and bonding techniques.

fibres, which resulted in non-wovens with higher areal density compared to needle-punched non-wovens. This difference leads to the bulkiness and densification achieved through chemical bonding, which enhances thermal resistance.

Biodegradation studies revealed that while banana fibres in the non-wovens degraded within approximately five months, the r-PET fibres remained intact and could potentially be recycled. This dual characteristic of biodegradability and recyclability underscores the environmental benefits of using natural fibres like banana alongside recycled synthetics in non-woven applications. The study thus supports the development of sustainable materials that align with global efforts towards environmental stewardship.

6.2.5.3 Case Study 3: Thermal Insulation Non-woven Using Kapok, Polypropylene, and Chicken Feather Fibre

Landfills are causing environmental pollution as a result of the disposal of waste with an abundance of natural fibres. Recycling these waste materials for uses like thermal insulation offers a viable way to address this problem and provide an environmentally responsible substitute. One vital use of thermal insulation is in cold-weather protective clothing, typically made through layering and filling techniques. A very popular and reasonably priced natural fibre, kapok, is distinguished by its sustainability, low weight, and special hollow lumen structure, which makes it perfect for thermal insulation. Furthermore, many studies have examined the recycling and repurposing of chicken feathers for clothing because they are inexpensive, readily available, and biodegradable. In this study, kapok and polypropylene are blended in two different areal densities and blend ratios to create needle-punched non-woven fabrics (7). Chicken feather fibres are then used as fillers in quilted composite structures (QCS).

Non-woven structures with higher kapok content demonstrated greater thermal resistance. Moreover, increasing the weight of chicken feathers from 300 to 800 g/m^2 improved the thermal resistance from 0.617 to 0.665°C m^2/W (see Figure 6.4). This enhancement is primarily due to the hollow, honeycomb-like structure of chicken feathers, which creates air pockets that trap heat, limiting its transfer to adjacent layers. The barb structure of chicken feathers, along with their low density, availability, and cost-effectiveness, makes them suitable for various applications, such as composites and outerwear. Despite these advantages, extracting chicken feather barbs from the rachis presents challenges due to the absence of standardized methods. However, their superior cohesiveness, due to their twisting and bending properties, makes them a promising material for sustainable textile products, potentially replacing conventional commercial products.

6.2.5.4 Case Study 4: Thermal Insulation Non-woven Using Natural Fibre Waste

This research focuses on exploring the potential of using natural fibre waste to create environmentally friendly and cost-effective insulation materials. The non-woven fabrics are developed from textile waste using the needle-punched technique for thermal insulation applications (8). This study focused on the utilization of three types of textile waste: CN, silk cocoon waste, and polyester/cotton flat strip waste. Four types of non-wovens are developed: 100% CN, 100% silk cocoon waste, polyester/cotton

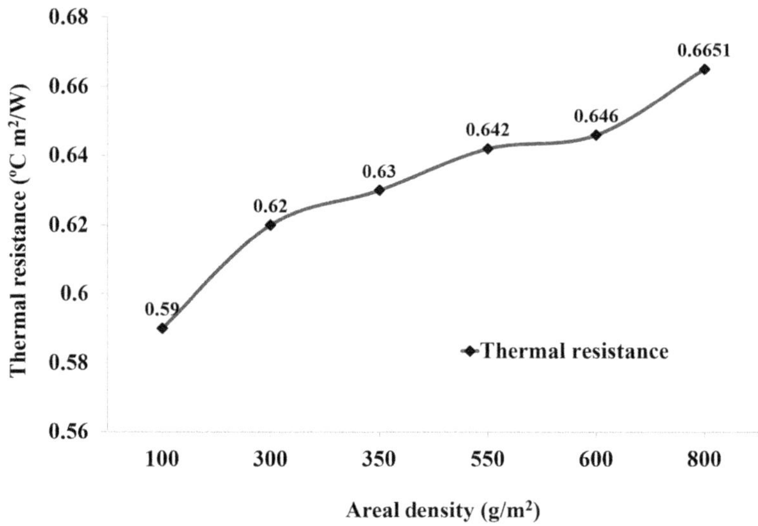

FIGURE 6.4 Thermal insulation properties of chicken feathers of different areal density.

flat strip waste (PCW), and silk cocoon waste/comber noil (SWCN), and are tested for thermal insulation performance.

The findings revealed that thermal resistance rose as the areal density and thickness of the non-wovens increased. Non-woven fabrics contain significant void spaces that trap air, thereby improving their insulation. Silk cocoon waste exhibited higher thermal insulation (0.201 m^2 K/W) than CN and PCW. Blending SWCN improved the thermal insulation (0.212 m^2 K/W) due to increased porosity created by CN since they can preserve more still air. The study compared the developed SWCN with a commercial thermal insulation product made of 100% polyester fibres. Despite being thinner, the developed non-woven demonstrated equivalent thermal resistance (0.212 m^2 K/W) compared to the commercial product (0.022 m^2 K/W) having a thickness of 10 mm. This indicates that the developed non-wovens are effective insulators and could be viable alternatives to conventional insulation materials.

6.3 ACOUSTIC INSULATION

6.3.1 Generation of Sound Waves

Sound is a form of energy that is produced by vibrating objects and propagated as mechanical waves through a medium like air, water, or solids. These vibrations are transformed into electrical signals that our brain interprets as sound. Sound is generated when an object vibrates and causes a disturbance in a surrounding medium (like air). These vibrations move outwards in all directions in the form of compressions and rarefactions, creating longitudinal waves. Compressions are areas where air molecules are densely packed, whereas rarefactions are areas where air molecules are more spread out. For instance, when a guitar string is plucked, the string vibrates, causing the surrounding air to compress and expand. This chain reaction of vibrating air molecules reaches the listener's ears, and the brain interprets the vibrations as sound. Any undesirable sound encountered by humans is termed "noise and noise pollution," and its ill effects are a hotly debated issue persisting in the 21st century.

A sound wave can be represented as a sine wave due to the successive form of compressions and rarefactions. A sine wave helps in visualizing the behaviour of sound waves over time, showing how the pressure or displacement of particles oscillates. **Wavelength** refers to the distance between two consecutive points that are in phase, such as two successive compressions or rarefactions in a sound wave. **Frequency** is the number of oscillations or cycles a sound wave completes in 1 s, expressed in Hertz (Hz). One Hertz is equal to one cycle per second. Higher frequency sounds correspond to higher-pitched tones (such as a whistle), while lower frequency sounds correspond to deeper, bass-like sounds (such as a drum). Frequency is inversely proportional to wavelength, meaning the higher the frequency, the shorter the wavelength. **Amplitude** is the maximum displacement of particles in a sound wave from their rest position. It determines the loudness of the sound, with greater amplitude leading to louder sounds. Amplitude is measured in terms of pressure differences in decibels (dB), and larger amplitudes correspond to higher energy and louder sound levels.

6.3.2 Physics Behind Acoustics

Acoustics is the branch of physics that focuses on the study of sound, specifically mechanical waves travelling through gases, liquids, and solids. It deals with the generation, control, transmission, reception, and impact of sound. Sound insulation refers to the materials or methods used to prevent the transmission of sound between spaces. The effectiveness of sound insulation depends on the material's ability to absorb, reflect, or dampen sound waves. Sound insulation helps in controlling noise by restricting the passage of sound from one area to another, improving acoustic comfort in buildings and preventing unwanted noise from entering a space. The ISO 25267 standard recommends that sound pressure levels should not exceed 30 dB in the living room, 40 dB in the kitchen, and 20 dB in the bedroom. Prolonged exposure to noise levels of 85 dB or above can cause elevated blood pressure, increased stress, and permanent hearing damage (19, 20). To maintain the noise level within the desirable limit, appropriate noise suppressors and insulators are used. The key criteria for selecting an insulating material include effective thermal and sound insulation properties. Typically, conventional insulating materials should have a noise reduction coefficient (NRC) between 0.5 and 0.9, and a thermal conductivity coefficient ranging from 0.030 to 0.045 W/m K (21).

6.3.3 Types of Sound Absorbers

Porous materials like fibreglass or foam work by absorbing sound energy, especially at high frequencies. As sound waves enter the porous structure, they encounter resistance, causing friction that converts the sound energy into heat. This mechanism helps to reduce noise by attenuating high-frequency sound waves. **Perforated thin materials** often work best for low-frequency sound. They contain holes or slits that allow sound waves to enter, creating resonance within the material. As the sound

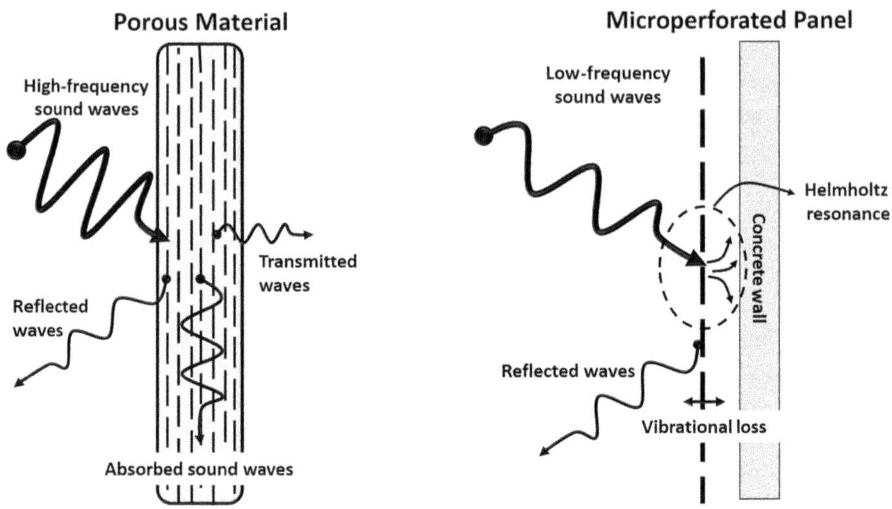

FIGURE 6.5 Mechanisms of sound absorbers.

waves interact with the cavities, they are dampened, reducing their intensity by the Helmholtz resonance principle (see Figure 6.5).

6.3.4 Factors Affecting Sound Insulation Properties

Non-woven materials are well suited for sound insulation and absorption applications because their fibrous structure and large surface area effectively help reduce environmental noise pollution (23). Non-woven materials perform differently across the sound spectrum. Typically, high-frequency sounds (e.g., human speech, machine noise) are absorbed more easily by non-woven materials due to their ability to interact with the finer pores and fibres. Low-frequency sounds (e.g., bass from music or heavy machinery) are harder to absorb and require thicker or denser non-woven materials. Areal density (mass), pore size, pore volume, tortuosity, bulk density, and thickness are important physical properties of non-woven fabrics that influence acoustic performance.

- Denser materials usually have a higher mass per unit area, which increases the material's ability to block sound transmission. However, if the material is too dense, it might reflect sound rather than absorb it, reducing its sound absorption effectiveness at certain frequencies.
- Thicker non-woven materials provide better sound insulation. Increased thickness adds more mass and layers, which improves the ability of the material to absorb and dampen sound waves, particularly at lower frequencies.
- Finer fibres generally create more surface area for sound waves to interact with, improving absorption. Additionally, the arrangement of fibres (random or aligned) affects how well the material can absorb or block sound.

- Materials with higher porosity tend to absorb more sound, especially high-frequency sounds, because sound waves enter the pores and lose energy through friction as they travel through the material.
- Non-woven materials with optimal airflow resistivity can absorb sound effectively by converting sound energy into heat. If the resistivity is too low, sound passes through the material without much absorption; if it is too high, the material reflects sound rather than absorbing it.

6.3.5 Use of Textile Waste in Sound Insulation Application

The applications of textile waste in sound insulation are diverse, spanning various industries such as construction, automotive, and consumer goods. Textile waste materials can be used to manufacture acoustic panels and tiles designed to absorb sound and reduce reverberation in open spaces like offices, public buildings, and entertainment venues. Companies that incorporate textile waste materials into their sound insulation products can differentiate themselves in the market by offering sustainable and innovative solutions. As industries continue to prioritize sustainability, the integration of textile waste into sound insulation products will play a crucial role in achieving environmental goals and improving quality of life through better acoustic comfort and noise reduction.

6.3.5.1 Case Study 1: Sound Absorption Panel Using Coir Mat

Coir, a natural fibre extracted from coconut husks, has the highest lignin content among all natural fibres (46–48%), which, combined with its low density, gives it significant porosity, making it ideal for sound absorption. According to literature, each coir fibre consists of multiple elementary fibres with lumens and a central canal (lacuna), resulting in 22–30% porosity, particularly effective for high-frequency sound absorption. In this study, the sound absorption behaviour of multilayer hybrid coir structures is examined (see Figure 6.6). A coir non-woven material, 24 mm thick, is compressed and arranged in a density gradient (DG24), with the lowest density layer facing the sound source to improve absorption at mid and high frequencies (9). To enhance low-frequency absorption, a hybrid face layer of compressed coir is sprayed with shredded denim waste fibres at varying proportions (0%, 5%, and 10%).

The DG24 structure showed strong absorption in the mid- and high-frequency range, with a sound absorption coefficient of 0.3 at 1,000 Hz, peaking at 0.76 at 2,000 Hz. The low-density face allowed sound to enter the structure easily, and as the sound travelled through denser layers, frictional losses converted sound energy into thermal energy. The higher density layers increased flow resistance and tortuosity, improving sound attenuation. This gradient structure performed better than uniform layers of the same thickness but lacked effectiveness in absorbing low-frequency sound. To address this, the addition of denim fibres on the sound face acted like a perforated panel, absorbing low frequencies while reducing high-frequency absorption. Adding 5% denim snippets doubled the NRC compared to plain coir. This research highlights the potential of using coir and waste denim in hybrid multilayer structures for broad-frequency sound absorption. The innovative design enhances

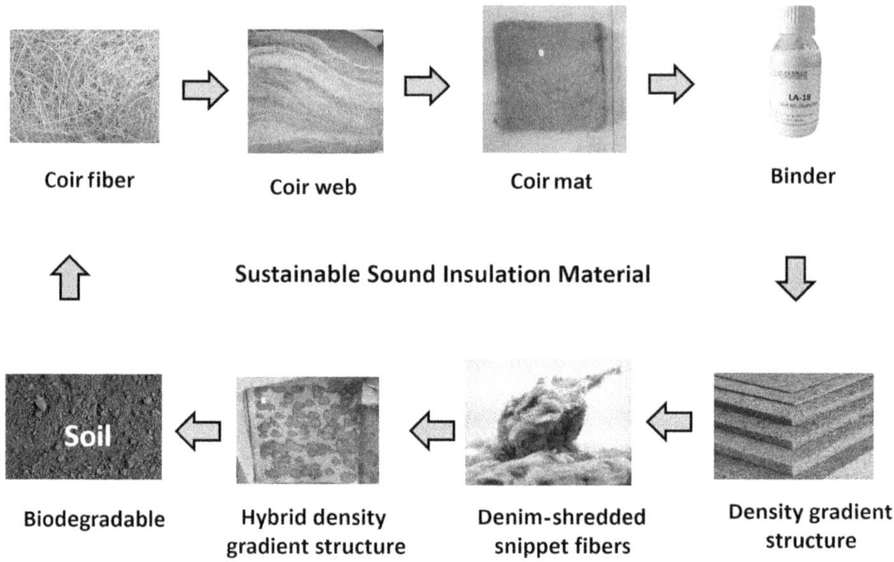

FIGURE 6.6 Process diagram of developing hybrid density gradient structure.

sound absorption, combining aesthetics because of the surface made of denim fibres and offering an eco-friendly solution to repurpose coir fibres and denim waste, making it practical for manufacturing and various applications.

6.3.5.2 Case Study 2: Sound Absorption Panel Using Non-woven Coir Mats

The quality of musical performances in concert halls greatly depends on their acoustic properties, including volume, surface area, and the absorption coefficient of various surfaces. Porous absorbing materials are widely used for sound absorption and are especially effective at attenuating medium- to high-frequency sounds. Resonance absorbers like perforated plasterboard and metal sheets are better for low frequencies. However, single structures often fail to cover the entire frequency range, necessitating sandwich structures that combine various materials for broad-spectrum sound absorption. In recent years, multilayer acoustic absorbers combining perforated plates and porous materials with airspace have been developed to absorb broadband noise, with their performance largely dependent on the manufacturing process.

This work aimed to develop hybrid fibrous panels from porous non-woven coir mats and perforated panels for broad-frequency sound control (10). Various configurations are tested, including porous non-woven coir mats alone, perforated panels with air gaps, and combinations of perforated panels with porous coir mats and air cavities (see Figure 6.7). Porous non-woven mats are created using a needle-punching machine and natural rubber polymer latex for structural integration. Sound absorption characteristics of these porous samples showed that porous materials generally have minimal sound absorption at low frequencies (<1,000 Hz) but perform well at higher frequencies. The maximum sound absorption coefficient observed is 0.69 at higher frequencies, with NRC values ranging from 0.08 to 0.18.

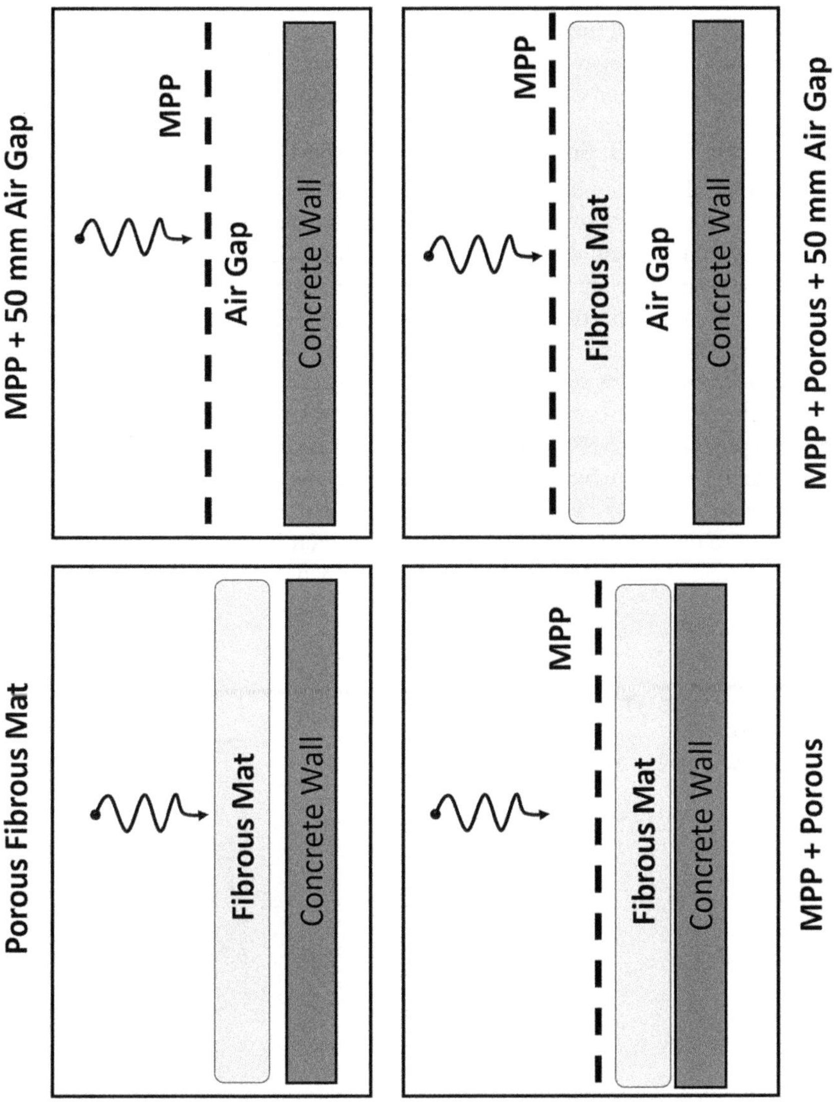

FIGURE 6.7 Arrangements of micro-perforated panel, porous non-woven, and air gap for sound insulation.

A micro-perforated panel (MPP) backed by a 50 mm air gap absorbed sound well in the low-frequency range but not at higher frequencies. When a porous coir mat is added between the MPP and a rigid wall, the sound absorption characteristics improved across both low and high frequencies, effectively creating wide-band absorbers. This hybrid structure could be fine-tuned by adjusting parameters like the thickness and density of the porous material, making these panels suitable for use in auditoriums and conference halls. This innovation of utilizing agricultural waste into effective sound insulation offers a sustainable solution for reducing waste.

6.3.5.3 Case Study 3: Sound Absorption Properties of Luffa Fibrous Mats

In recent times, there is an increasing interest in developing alternative sound-absorbing materials from natural waste fibres, due to the harmful effects caused by synthetic fibres. In accordance with that, one promising fibre is *Luffa cylindrica*, a fruit-based fibre derived from the cucumber family, often used as bath sponges. These fibres are generally available in fibrous mat form. The surface nature of luffa fibres contains hollow, punctuated cords with a low specific density of 1.48 g/cm^3. Due to its hollow structure, luffa mats show relatively low NRC. However, researchers found that introducing a layer of kapok-cotton web to the luffa mat significantly enhanced its sound absorption properties (see Figure 6.8). This improvement is primarily due to the large lumen of kapok fibres, which is its unique characteristic among other natural fibres. This lumen space creates a more tortuous path for sound waves, increasing airflow resistance. The lightweight nature of kapok fibres increases the fibre vibration caused by sound waves, resulting in momentum loss and better sound absorption (11).

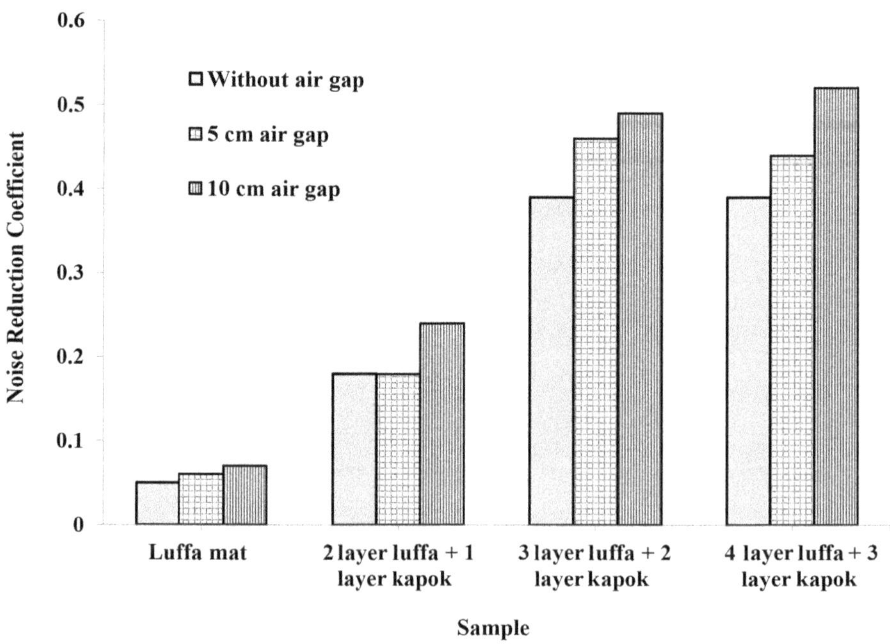

FIGURE 6.8 Noise reduction coefficient of different layers of luffa and kapok fibres.

Further investigations revealed that increasing the thickness of the luffa mats from 20.13 to 25.6 mm resulted in a 130% improvement in NRC. This enhancement is attributed to the higher porosity in thicker mats, which causes greater frictional losses and more effective sound energy damping. The addition of an air gap behind the sample during testing further improved the NRC, demonstrating the importance of configuration in sound absorption performance. Though luffa fibrous mats alone do not exhibit excellent NRC values, their combination with kapok-cotton web and adjustments in thickness and testing conditions significantly enhance their performance. As eco-friendly, non-toxic materials with fairly good sound absorption capabilities, luffa-based mats show great potential for use in acoustic applications in buildings and automobiles.

6.3.5.4 Case Study 4: Sound Absorption Non-wovens Using Natural Fibres

In the pursuit of eco-friendly solutions for automotive interior noise reduction, this research has explored the development of renewable and biodegradable non-woven materials made from natural fibres like banana, bamboo, and jute. Traditional automotive interior components are often made from materials such as glass fibres and synthetic foams, which are difficult to recycle. To address this, three types of non-woven materials are created using a needle-punching technique, blending natural fibres with polypropylene staple fibres in 50:50 ratios, and tested for their sound absorption properties. The study also compared various physical and comfort properties of the developed materials, including areal density, thickness, stiffness, tensile strength, elongation, air permeability, and thermal conductivity (12).

A number of important metrics showed that the bamboo/polypropylene non-woven performed better than the other materials that are developed. Compared to the banana and jute fibre blends, it showed reduced elongation, air permeability, and thermal conductivity, but its compact structure helped to increase its tensile strength and stiffness. An excellent option for reducing noise inside cars is the bamboo/polypropylene blend, which also demonstrated a high sound absorption coefficient. The sound absorption coefficients of the bamboo/polypropylene and jute/polypropylene non-wovens reached the desired levels at a frequency of 800 Hz. At this frequency, the banana/polypropylene blend, however, performed 22% worse than expected.

Nevertheless, at higher frequencies (1,600 Hz), all of the non-wovens—including the bamboo blend—exhibited decreased sound absorption, suggesting that more advancements could be made by thickening the non-woven materials. Because of its excellent sound-absorbing qualities, stiffness, and high tensile strength, the bamboo/polypropylene blend turned out to be the best choice. Using renewable and biodegradable materials like jute, banana, and bamboo has several environmental advantages, which makes these non-woven composites appealing substitutes for conventional car interior materials. The creation of these natural fibre non-wovens is a major step towards developing environmentally friendly solutions for vehicle noise control, which is in line with the automotive industry's increasing demand for sustainable products. They could perform better at higher frequencies with additional optimization like changing the material's thickness, which would increase their usefulness for a larger variety of acoustic applications.

6.4 OIL SORBENT

6.4.1 Basic Principles of Sorption

One substance adheres to or is absorbed into another through the physical and chemical process of sorption. Interaction between the sorbent (substance that absorbs or adsorbs another substance) and the sorbate (substance that is being absorbed or adsorbed by the sorbent) is called sorption. These interactions are influenced by intermolecular forces (such as van der Waals forces, hydrogen bonds, and ionic bonds) and surface chemistry. The basic principle of sorption in the context of oil sorbents involves the interaction between the sorbent material and oil, where the material either absorbs or adsorbs the oil. These two mechanisms—absorption and adsorption—play a crucial role in the effectiveness of oil sorbents for cleaning up oil spills. **Adsorption** is a process in which one substance adheres to the other substance's surface and does not enter inside the substance. Adsorption is assisted by physical sorption (weak van der Waals forces) and chemisorption (formation of chemical bonds). **Absorption** is a process in which a substance is taken into another substance. This mechanism is assisted by volume diffusion (oil penetrates the bulk material) and surface diffusion (oil moves along the surface of the sorbent).

6.4.2 Types of Oil Sorbents

Sorbents can be either natural or synthetic. Natural sorbents are categorized into organic types, such as peat moss or wood-based products, and inorganic types, including vermiculite and clay. Recently, synthetic sorbents have seen increased use in oil spill cleanup efforts. These synthetic materials are often employed to wipe down other cleanup tools like skimmers and booms after an oil spill response. The effectiveness of a sorbent depends on its surface area available for oil adhesion as well as the nature of its surface.

A fine, porous sorbent with numerous small capillaries offers a large surface area and is most suitable for recovering light crude oils or fuels. In contrast, sorbents with a rougher texture are better for cleaning heavier oils like crude oil or bunker fuel. Sorption occurs through two simultaneous processes: absorption and adsorption. Absorption involves one substance permeating the internal space or volume of another, whereas adsorption takes place on the surface of a material. In adsorption, intermolecular forces cause molecules to adhere to the surface, while absorption involves the soaking in of a liquid or gas by a solid without such molecular forces (27).

6.4.3 Features of Oil Sorbent

Oil sorbents are materials that absorb or adsorb oil, effectively removing it from water. The effectiveness of an oil sorbent is influenced by various physical properties of oil, including surface tension, density, and viscosity, represented in Table 6.1.

The basic physics underlying oil sorbents, with a focus on surface tension and related phenomena, are as follows.

TABLE 6.1
Physical Properties of Oil at 20°C

Properties	High-Density Oil	Diesel Oil
Surface tension, mN/m	31	25
Density, kg/m^3	900	820
Kinematic viscosity, mm^2/s	130	2

6.4.3.1 Surface Tension

Surface tension is the cohesive force present at a liquid's surface that causes it to act like a stretched, elastic membrane. It arises due to the attractive forces between the molecules at the surface. For oil sorbents, surface tension is a critical property because it influences how oil spreads on water and how the sorbent interacts with the oil. The higher the oil's surface tension and viscosity, the more ideal the oil spill conditions.

- **Oil-water interface:** Water and oil do not mix because they are immiscible liquids. The oil forms thin layers or droplets on the water's surface due to surface tension at the oil-water interface. The sorbent's ability to absorb or adsorb oil is directly related to how well it can interact with these oil droplets.
- **Wettability:** The ease of a liquid spreading across a solid surface is known as wettability. An oil sorbent needs to be both hydrophobic (repelling water) and oleophilic (attracting oil) in order to work. This ensures that the sorbent attracts and absorbs oil while repelling water. Surface tension plays a key role in determining the wettability of the sorbent.

6.4.3.2 Absorption and Adsorption

Oil sorbents can remove oil from water through two main mechanisms: absorption and adsorption. Similar to how a sponge absorbs water, the sorbent absorbs oil into its internal structure during this process. Since this process is bulk, the oil is dispersed throughout the substance. The porosity of the sorbent and the oil's viscosity determine absorption efficiency. Conversely, adsorption is the process by which oil sticks to the sorbent's surface. This is a surface phenomenon, and the sorbent material's surface area and characteristics determine how well adsorption works.

6.4.3.3 Role of Capillary Action

Capillary action is a process of liquid entering a small space without the help of any external forces. This makes it easier for the oil to move through a sorbent's pores and spaces, which increases its capacity to absorb more oil. The interplay of surface tension and adhesive forces between the oil and the sorbent material drives this process.

6.4.3.4 Material Properties of Oil Sorbents

Oil sorbents must have properties such as porosity, a large surface area, and both hydrophobic and oleophilic characteristics to effectively remove oil, especially from water. These features help the sorbent absorb more oil while repelling water. Some common sorbents include natural fibres like cotton, inorganic materials such as clay and silica, and synthetic polymers like polypropylene. For water surfaces, sorbents need to be lightweight, oil-attracting, environmentally safe, and easy to handle. Their performance can be improved through fibre blending, surface modification, or chemical treatments. Combining natural and synthetic fibres enhances oil absorption and durability by taking advantage of the strengths of each material.

6.4.4 Process Sequence of Oil Sorbent in Oil Spill

There is a sequence involved in using an oil sorbent to absorb oil and then releasing the oil through squeezing (see Figure 6.9).

- **Deployment of oil sorbent:** The sorbent material (e.g., pads, booms, or loose sorbent material) is placed directly on or near the oil spill, either on the water surface or on land. In water, sorbent booms are often used to contain and absorb the oil.
- **Absorption or adsorption of oil:** The oil sorbent begins to interact with the oil spill either by adsorption or absorption. The sorbent continues until it reaches its maximum capacity. At this point, the material will become fully saturated with oil.
- **Recovery of the oil-soaked sorbent:** Once the sorbent has absorbed enough oil, it is carefully retrieved from the spill site. In water environments, this often involves using booms or nets to collect oil-soaked pads or booms.
- **Extraction of oil:** Mechanical processes such as squeezing or wringing are applied to the sorbent material to release the oil it has absorbed. This can be done manually (for small spills) or using specialized equipment like hydraulic presses (for larger-scale operations).

Depending on the type of oil, it may be possible to recover and reuse the oil after filtering and treating it. Some sorbent materials are designed to be **reusable**. After squeezing, if the material remains intact and has retained its absorptive properties, it may be redeployed to absorb more oil. Reusable sorbents, such as polypropylene-based materials, can often go through several cycles of absorption and oil extraction before they are discarded. Once the sorbent has reached its maximum usage capacity or is no longer effective, it is collected for disposal (landfill or incineration). It is typically beneficial if the sorbent is biodegradable or can be safely disposed of by incineration without generating excessive black smoke or harmful emissions. To utilize textile waste for oil sorption, the waste material is modified to enhance its oil-absorbing properties.

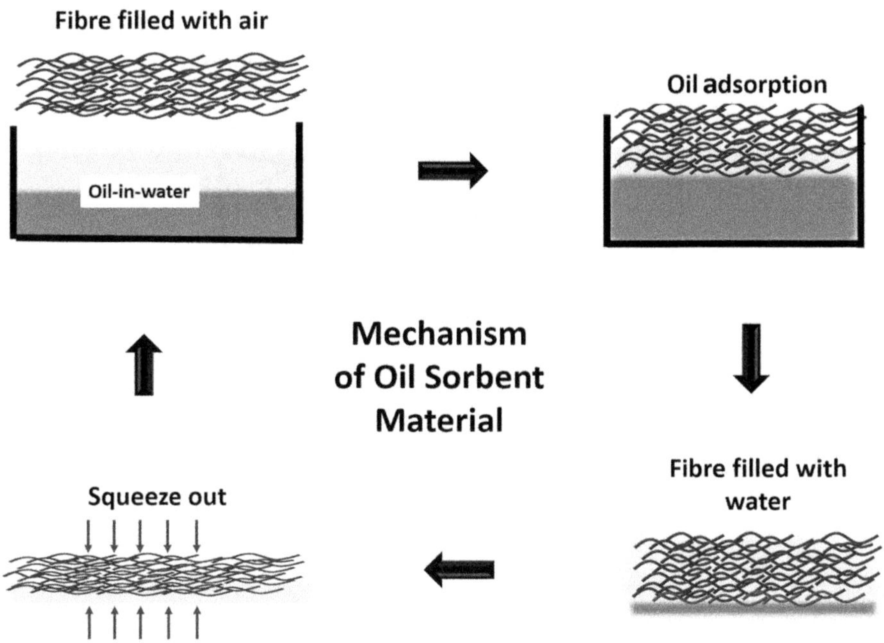

FIGURE 6.9 Process sequence of oil sorbent in oil spill.

6.4.5 Measurement of Oil Sorption Capacity of Fibrous Assemblies and Non-wovens

To measure the oil sorption behaviour of textile fibres, the fibres are packed inside a circular PVC tube with a known diameter and length (see Figure 6.10). The experiment aimed to achieve a specific porosity within the tube by filling it with a known volume of fibres. Holes, each 1 mm in diameter, are drilled at equal intervals around the tube to allow liquids to flow freely into the interior. The porosity was adjusted to range between 91% and 99%, calculated using the formula:

$$\text{Porosity} = \left(1 - \frac{m_f}{\rho_f - V_{mf}}\right) \times 100$$

Where m_f is the mass of the fibres (kg), ρ_f is the density of the fibres (kg/m^3), and V_{mf} is the volume of the PVC tube (m^3). The appropriate mass of fibres is placed inside the tube to achieve the desired porosity. When the fibre-filled tubes are ready, they are submerged in an oil bath for five minutes or until they are completely saturated. To enable any extra oil on the surface to drip off, the tubes are taken out of the immersion and held horizontally for 30 s. A 1 g sample is immersed in the oil for 20 min in a similar process for non-woven fibrous assemblies. It is held horizontally for 30 s

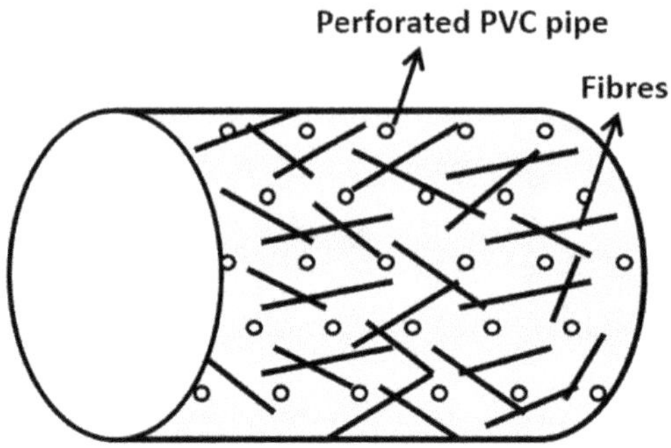

FIGURE 6.10 Perforated PVC tube filled with fibres.

to remove the surface oil. The oil sorption capacity of the fibres is then determined using the following equation:

$$\text{Scorption}(g/g) = \frac{w_f - w_i}{w_i}$$

Where w_f is the final weight (g), and w_i is the initial weight of the material (g).

6.4.6 Sorption Isotherm

The relationship between the amount of material adsorbed or absorbed by a material is described by a sorption isotherm, which is a graphical representation of the sorption behaviour. Adsorption and absorption isotherms are two common forms of sorption isotherms. A curve that illustrates how the amount of a material adsorbed on a surface changes with its concentration in the surrounding phase at a constant temperature is called an adsorption isotherm. The two most well-known models are the Freundlich and Langmuir isotherm theories. These models help predict the amount of material that will be adsorbed by taking into account variables like surface area and concentration. The amount of the absorbed material that is absorbed by the bulk material at various concentrations is displayed by an absorption isotherm at constant temperature. To explain this, for instance, Henry's law or the BET model for gas absorption in porous solids are commonly employed.

6.4.7 Application of Textile Waste as Oil Sorbent

In order to clean up oil spills in rivers, lakes, and oceans, textile waste can be converted into efficient oil sorption materials. These substances are helpful because they repel water and absorb oil, which speeds up and improves cleanup. Moreover, they are effective at eliminating oil from polluted soil and industrial locations. This fibrous sorbent helps in extraction and reuse of the absorbed oil. Value can be added to the material by converting the leftover materials into beneficial products like biofuels, biocomposites, or insulation after oil recovery. In addition to helping manage textile waste, this circular approach conserves resources by repurposing waste.

6.4.7.1 Case Study 1: Oil Sorbent Non-woven Using Cotton Comber Noil and Recycled Polyester

Cotton comber noil (CCN) is a valuable by-product from the spinning industry which is used to spin open-end yarns with a coarser texture. Blending CCN with recycled polyester (r-PET) fibres to develop a sustainable non-woven fabric for oil spill cleanup is investigated in this study. The oil sorption property of the developed non-woven for both crude and vegetable oil is evaluated using the ASTM F 716-09 standard. According to the study, the non-woven development's ability to absorb oil is improved by raising the blend proportion of CCN (13).

The effectiveness of oil sorption is influenced by multiple factors. Cotton fibres, for instance, naturally contain surface waxes that facilitate oil adsorption through C–H bond interactions between molecules. Furthermore, cotton fibres' physical characteristics—such as their porous lumen—allow oils to be retained and diffused, which increases absorption even more. Cotton's fineness is a result of its low micronaire value, which increases the surface area available for capillary action and oil adsorption, improving the fibre's ability to absorb oils. However, because of its oleophilic-hydrophobic characteristics, r-PET promotes oil absorption, and its functional groups help to speed up oil uptake. The combined CCN/r-PET non-woven fabric's hydrophobic and oleophilic properties enhance the process in a matter of seconds by exhibiting high oil sorption. However, repeated use reduces the fabric's sorption capacity, mostly because of structural deformation and partially extracted residual oil.

Over time, the needle-punched non-woven fabric composed of CCN/r-PET loses its structure and shape due to oil absorption, which impairs its functionality. Additionally, some of the oil becomes lodged in the fibres, reducing their capacity to absorb additional oil in the future. The fabric is still reusable despite this performance decline. It can still absorb about 18–3 g of crude oil and 13–5 g of vegetable oil per gram of fabric after five uses. This demonstrates that even though its effectiveness declines, the fabric is still beneficial for several cleanups.

6.4.7.2 Case Study 2: Oil Sorbent Non-woven Using Silkworm Cocoon Waste

Waste from silkworm cocoons (SCW), particularly the inferior cocoons that are typically thrown away by the industry, has a lot of potential as an inexpensive environmentally friendly raw material for oil sorbents. Due to their accessibility and frequent disregard, these cocoons are an underutilized resource. Silk is appropriate for oil-water separation because it naturally repels water and attracts oil. In this case study, the production of needle-punched non-woven fabrics using SCW is investigated along with the material's capacity to absorb oil, be recycled, and decompose naturally over time. As shown in Figure 6.11, the study emphasizes the potential of silkworm cocoon waste (SCW) as an environmentally friendly and low-cost material for oil sorbent production. This is especially true of the second-quality cocoons that are typically thrown away by the industry. Because they are so accessible and frequently disregarded, these cocoons represent an unexplored resource. Silk can be used to separate oil from water because it naturally repels water and attracts oil. In this case study, the production of needle-punched non-woven fabrics using SCW is investigated along with the material's capacity to absorb oil, be recycled, and

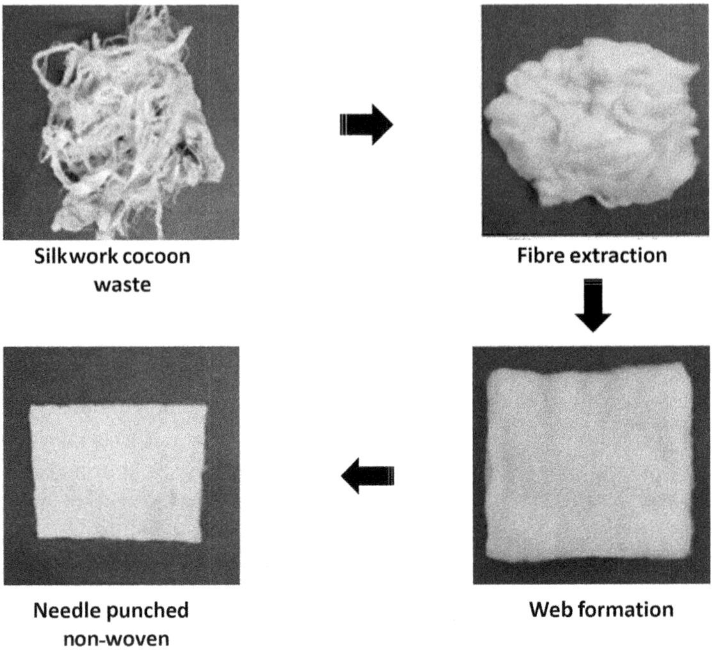

FIGURE 6.11 Development process of needle-punched non-woven from silk cocoon waste.

decompose naturally over time. The study emphasizes SCW's potential as an environmentally friendly oil spill cleanup method (14).

The oil sorption results showed that fabrics with the highest porosity has the highest oil sorption capacities for crude oil and vegetable oil, which are about 31.52 g/g and 25.92 g/g, respectively. The higher viscosity of crude oil compared to vegetable oil contributed to its greater sorption capacity. The oil retention properties of the non-woven fabrics are in the range of 92–95% for crude oil and 88–91% for vegetable oil. The variation in retention abilities is attributed to the different viscosities and adhesiveness of the oils. High-viscosity oils, like crude oil, tend to adhere more strongly to the fibrous substrate, resulting in better retention. In the contact angle test, the water droplets remained nearly spherical on the fabric surface, indicating a high water contact angle of 123°, while oil droplets rapidly penetrated the fabric, showing zero oil contact angle. These results confirmed the hydrophobic and oleophilic nature of the silk non-woven fabrics, essential characteristics for effective oil sorbents.

The reusability of the silk non-woven fabrics is tested over five sorption-desorption cycles. The oil sorption capacity decreased gradually with each cycle, with final capacities of 13.5 g/g for crude oil and 10.3 g/g for vegetable oil after five cycles. Irreversible deformation of the fibre structure, tightening of interfibre spaces, and residual oil in the fabrics are some of the reasons for the performance drop. Notwithstanding this decrease, the textiles showed good reusability, which made them economical for managing oil pollution. From the results of the soil burial test, it is observed that the fabrics are entirely decomposed in 100 days, proving to be environmentally friendly. This rapid degradation indicated that the materials are biodegradable and safe to discard as landfill after use.

The fabrics possess both oleophilic and hydrophobic properties, which enable them to absorb oil, and their biodegradability guarantees that they have minimal environmental impact after use. This approach not only provides an effective way to manage oil spills but also encourages the use of industrial waste (second-quality silkworm cocoons), which is often discarded as landfill. This strategy encourages a circular economy by repurposing waste materials to create useful environmentally responsible goods. The study's silk non-woven fabrics demonstrated promise for commercial use in environmental cleanup, especially for the removal of oil spills, because of their sustainability and performance advantages.

6.4.7.3 Case Study 3: Oil Sorbent Non-woven Using Recycled Polyester

Over 75% of the world's production of synthetic fibre is polyester, making it the market leader. A large amount of recycled polyethylene terephthalate, or r-PET, is used to make polyester fibres, making it the most recycled plastic in the world. r-PET fibre makes up a sizeable amount of recycled goods and is primarily made from post-consumer PET bottles using melt spinning technology. Studies have concentrated on the potential of r-PET fibres for a range of technical textile applications because synthetic fibres are linked to environmental issues. The goal of this project is to produce needle-punched non-woven textiles from recycled polyester fibres and assess how well they work in applications involving oil spill cleanup. Different needle-punch densities, needle penetration depths, and fabric areal weights are used to create 15 distinct needle-punched non-woven fabrics. Important parameters such as oil sorption capacity, oil retention, oil sorption kinetics, wettability, and reusability are then methodically examined for these textiles (15).

Up to 24–85 g of crude oil and 20–58 g of vegetable oil per gram of fabric are found to be absorbed by recycled non-woven polyester fabrics. For crude oil and vegetable oil, the oil retention rates are roughly 93–96% and 87–91%, respectively. These textiles demonstrated potent oleophilic and hydrophobic qualities, exhibiting a static contact angle of 0° against vegetable and crude oils and 138° against water. Even after several cycles, reusability tests showed that the recycled polyester non-woven fabric maintained a sizeable oil sorption capacity. While earlier research has looked at a variety of textile uses for recycled polyester fibres, there have not been many in-depth studies on the oil sorption properties of needle-punched non-woven fabrics made from these fibres. Important information from this study may help direct the creation of efficient materials for cleaning up oil spills.

It is clear from the study that needle-punched non-woven fabrics composed of recycled polyester fibres are appropriate for cleaning up oil spills because of their high oil sorption capacity, good oil retention, and substantial reusability. Products manufactured from r-PET fibres may perform similarly to those made from virgin polyester fibres. This study supports the creation of sustainable and efficient materials from recycled resources for oil spill mitigation by providing important data to the field.

6.4.7.4 Case Study 4: Oil Sorbent Using Cotton, Kapok, Milkweed, and Polypropylene Fibres

In this study, oil sorbent is developed by layering non-woven fabric made from natural fibres such as kapok, cotton, and milkweed, blended with synthetic polypropylene fibres, to determine their oil sorption and retention properties. In particular, kapok and milkweed fibres offer environmental benefits as they can grow in barren land with minimal water requirements and without the need for fertilizers. Generally, these kinds of plants regenerate continuously without generating waste like seasonal crops. Kapok and milkweed fibres are sustainable and eco-friendly alternatives to synthetic materials. This case study utilized three different varieties of milkweed, along with cotton and polypropylene, to develop thermally bonded non-woven fabrics. Due to the lack of cohesion among the kapok and milkweed fibres, traditional carding technology is not suitable for forming a uniform web. Instead, air-lay non-woven technology is employed to develop the fibre webs.

The study revealed that various material and geometric factors, such as fibre type, diameter, surface area, porosity, and blend ratios, significantly influenced the oil sorption behaviour of the thermally bonded non-wovens (16). From Figure 6.12, it is clear that polypropylene fibres exhibited higher contact angles with both high-density and diesel oil compared to milkweed, kapok, and cotton fibres, which showed lower values. This increased contact angle in polypropylene could be attributed to polar chemical groups introduced by the spin finish, which reduced its affinity for oils.

Fibre porosity and surface area are found to be key factors influencing the oil sorption and retention abilities of the non-woven fabrics. The highest oil absorption recorded is 40.16 g/g for high-density oil and 23.00 g/g for diesel, indicating that

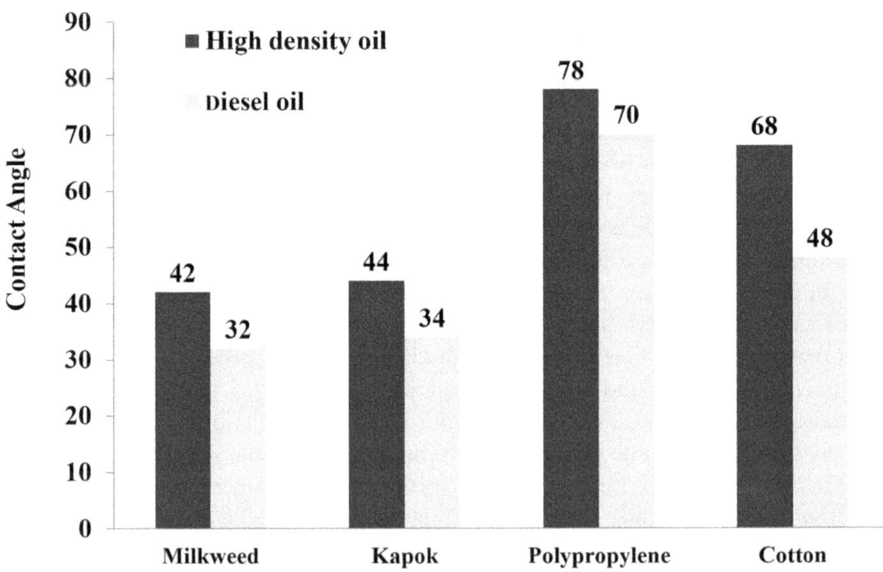

FIGURE 6.12 Contact angles of different fibres with high-density and diesel oil.

greater porosity and larger surface area significantly improve performance. Non-woven fabrics made from CG and CP fibres are especially efficient in soaking up oil without leaving any residue. Their buoyant nature, along with hydrophobic and oleophilic characteristics, makes them strong candidates to replace synthetic oil sorbents.

In conclusion, natural fibre-based non-wovens—especially those using kapok and milkweed fibres—showed strong potential for cleaning oil spills. Their excellent buoyancy and oil-attracting properties, combined with water-repelling behaviour, make them ideal for removing oil from water. As biodegradable materials, these fibres offer an eco-friendlier alternative. With high oil absorption, good retention, and sustainability, they stand out as promising replacements for synthetic materials in oil spill response.

6.4.7.5 Case Study 5: Oil Sorbent Non-woven Using Nettle Fibre

Polypropylene fibre-based oil sorbents are widely used for cleaning up oil spills. However, since polypropylene is non-biodegradable, it poses environmental pollution risks. Consequently, oil sorbents based on natural fibres have drawn great attention as environmentally friendly substitutes for synthetic materials. Several natural fibre materials, including bagasse, cotton, kapok, milkweed, and cattail fibre, have been investigated by researchers recently for the cleanup of oil spills. Of these, kapok fibres are particularly noteworthy because of their high capacity for oil sorption and biodegradability. Researchers think that kapok fibres could be a very efficient oil sorbent because of their uniform pore sizes and variable porosity. This study focuses on developing structured natural fibre non-wovens using a combination of nettle and kapok fibres (17).

The study investigated the ability of fibrous assemblies and needle-punched non-woven structures composed of 100% nettle and blends of kapok and nettle fibres in 50/50 and 75/25 ratios to sorb oil. The maximum oil sorption capacities of the nettle fibrous assemblies are 56 g/g for high-density oil and 23–9 g/g for diesel oil. The porosity of the assemblies varied between 91% and 99%. The findings unequivocally demonstrated that porosity significantly affected oil absorption because higher porosity resulted in higher oil retention. In comparison to non-wovens made entirely of nettle, kapok fibres enhanced oil sorption while decreasing water absorption when incorporated into the nettle-based structure. This improvement results from kapok fibres' inherent hydrophobicity. The structure of the fibres made it impossible to produce a 100% kapok needle-punched non-woven, so nettle is added to create a more stable fabric. Outperforming commercial polypropylene-based non-woven sorbents, the 50/50 nettle/kapok blend demonstrated the highest oil sorption capacity, absorbing 28.5 g/g of high-density oil and 22.5 g/g of diesel oil.

In conclusion, blending kapok fibres with nettle significantly enhanced the oil sorption capacity of nettle-based non-woven materials by 13–18% compared to pure nettle non-wovens. The study showed that both 100% nettle and nettle/kapok blends outperformed commercial polypropylene oil sorbents in oil absorption. From the results, the natural fibre non-woven pads have the potential to be a sustainable and eco-friendly alternative to synthetic options.

6.5 CONSTRUCTION

Using textile waste to create value-added products has also shown promise in the construction sector. Nowadays, the use of these materials in the field of construction has grown popular due to their special qualities and advantages for the environment. Thermal and acoustic insulators made from fibrous material, especially from textile waste, reduce the environmental impact of building projects. Additionally, these materials save building energy by maintaining ideal interior temperatures and lowering outdoor noise.

Fibre-reinforced composites are perfect for improving the mechanical qualities of composites because of their high tensile strength and lightweight nature. Waste textiles can be shaped into various forms to create creative architectural elements that are both aesthetically pleasing and environmentally advantageous. For instance, bricks, blocks, and panels made from textile waste can be used as environmentally friendly substitutes for conventional concrete materials. Additionally, the texture, colour, and design of these materials can be altered, giving architects and builders a wide range of choices for green building projects.

6.5.1 Case Study 1: Building Insulation Using Micro-dust Waste

There is a growing interest in developing insulation materials for building sectors which enhance performance along with sustainability. Recycling textile waste, particularly micro-dust from textile spinning mills, offers a potential solution for reducing landfill waste and environmental pollution. This study explores the use of micro-dust to create composite panels through compression moulding for applications in the construction sector, particularly as thermal barriers (see Figure 6.13). Ten composite samples with thicknesses of 6 and 10 mm are manufactured by adjusting the micro-dust-to-resin blend ratios, with resin content varying between 10% and 30% (18).

The mechanical properties of these composites, such as tensile, flexural, and compressive strength, along with fire retardancy, water absorption, and insulation properties, were tested. The results revealed that tensile strength decreased as fibre content increased, largely due to poor adhesion between the resin and fibres. Additionally, thicker 10 mm panels exhibited lower tensile strength compared to 6 mm panels because of their lower density and poor packing. Flexural strength also dropped with higher fibre content, as increased voids between fibres led to weaker interactions with the matrix, resulting in material failure during bending. Weak bonding and the presence of voids led to a decrease in mechanical strength as both density and resin content were reduced. However, none of the samples exhibited dripping, which is beneficial for building applications since synthetic foams often drip when exposed to fire. Additionally, all micro-dust-based composite panels demonstrated superior thermal resistance compared to commercial construction materials. Water absorption tests indicated that thicker composites, which contained more micro-cracks and gaps, absorbed more water, whereas thinner panels with smoother surfaces absorbed less due to fewer entry points for water.

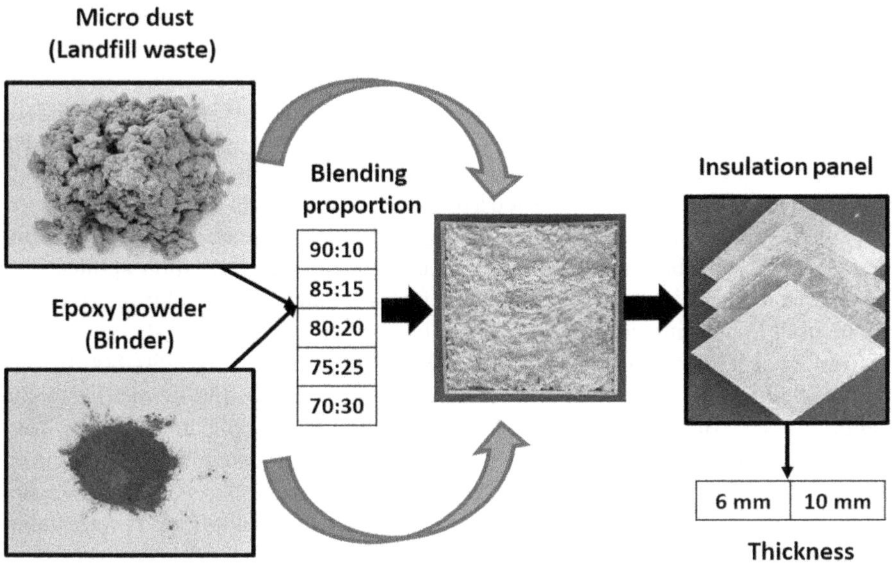

FIGURE 6.13 Development process of thermal insulation panel from micro-dust waste (19).

In conclusion, the 6 mm thick composite board with 70% micro-dust displayed the most desirable mechanical properties, including a tensile strength of 6.41 MPa, compressive strength of 3.6 MPa, and flexural strength of 10.2 MPa, all of which were superior to those of commercially available gypsum false ceiling boards of similar thickness and density. The thermal insulation value of the composite is 0.117 m^2 K/W, comparable to the commercial sample's 0.102 m^2 K/W. The composite panels developed in this study are not only comparable to commercially available building panels in terms of performance but also offer a cost-effective and sustainable alternative for thermal insulation.

6.5.2 Case Study 2: Building Insulation Panel Using Micro-dust and Coir Pith

Coir pith and cotton micro-dust fibres from the spinning industry are two underutilized materials which can be used for thermal insulation applications. Coir pith has specific properties such as low density, high porosity, and excellent water retention capacity, while cotton micro-dust contributes to environmental pollution when disposed of in landfills. The thermal insulation panels were manufactured using cotton micro-dust, coir pith, and epoxy resin, where adding coir pith to the panel composition greatly increased the panel's ability to withstand heat transfer (19).

Compression moulding technique is used to develop the composite panels at 150°C for 45 minutes under 150 bar pressure. This systematic approach ensured consistent panel quality and performance across all experimental samples. The study explored the thermal resistivity of the composite panels. Coir pith's low thermal conductivity and the large number of micro-voids in its fibrous structure, which further obstruct heat flow, are credited with this improvement. Consequently, the panels exhibited exceptional thermal insulation properties. When compared to standard commercial products, the *R*-values of the coir-enhanced panels are markedly higher, highlighting their potential as effective insulating materials in various applications. Due to voids and the fact that coir pith fibres are naturally weaker than cotton micro-dust, mechanical strength tests showed a decrease as the amount of coir pith increased. Even though the panels' mechanical strength slightly decreased as their coir pith content increased, they are still strong enough for thermal insulation applications. Because of their dual advantages of mechanical strength and thermal insulation, they can be used in building construction where both qualities are essential.

Compared with medium-density fibre (MDF) boards (density: 750–800 kg/m^3), the composite panels exhibited a lower density range of 500–730 kg/m^3. This lightweight property of the composite not only enhances their applicability in construction but also has advantages such as cost-effectiveness and sustainability. In addition, the manufacturing process is notably simple and efficient. It relies on a one-step compression moulding method that eliminates the need for complicated pretreatment or multiple processing stages. This straightforward approach helps reduce both production time and resource usage, contributing to overall cost-effectiveness. By adding value to waste streams such as cotton micro-dust and coir pith through innovative technical applications, the research promotes a circular economy approach in construction materials. This approach not only minimizes waste but also contributes to reducing the carbon footprint associated with traditional building materials.

6.5.3 Case Study 3: Building Insulation Materials from Textile Fibre Wastes

Textile waste fibres contribute massively to pollution, with millions of tons of fabrics thrown away or burned each year. The current study investigates the potential of textile fibre waste (recycled polyester, r-PET, CN, and silk cocoon waste) as sustainable thermal and sound insulation materials in the construction industry. The study reported herein presents results on the development of non-woven materials from these textile wastes and on their thermal insulation performance in buildings. Through textile waste recycling, the environmental burden of landfill is reduced, and at the same time new possibilities are created to produce cost-effective materials that have the potential to compete with established insulation products.

The research involved developing needle-punched non-wovens from 100% CN, 100% silk cocoon waste, and 100% recycled polyester fibres, as well as blends of these fibres (20). The thermal insulation performance, sound absorption, and mechanical properties of the non-wovens are measured. The non-wovens containing mixed fibres performed better than those with 100% of the single fibres, as seen in the results. For example, the 50% silk cocoon waste/50% r-PET non-woven fabric presented a higher thermal resistance due to the empty spaces that are filled by air, which further increases its insulation properties (see Figure 6.14). The thermal insulating performance and sound absorption coefficient of the silk cocoon/comber noil combination are also high, offering it as a possible alternative to common

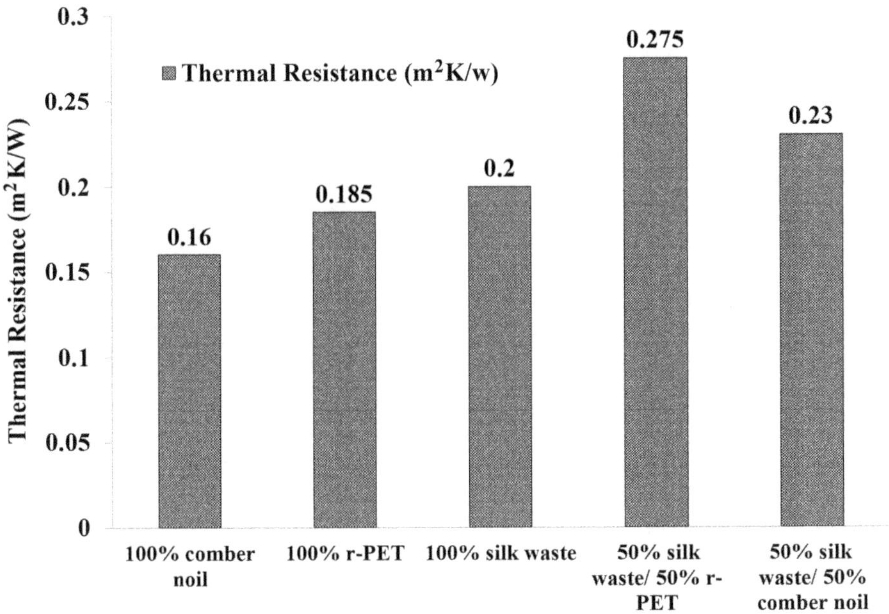

FIGURE 6.14 Thermal insulation properties of various textile waste.

insulation materials. It has good soundproof performance; the sound absorption can absorb sound in the air and reduce noise.

Moreover, the study found that these textile waste-based non-wovens outperformed some commercial insulation products in terms of thermal conductivity and sound absorption. For example, the developed non-wovens has a lower thermal conductivity than commercial polyester-based products, highlighting their efficiency in maintaining indoor temperatures. Furthermore, the biodegradability of natural fibres like silk and cotton is far more environmentally friendly than synthetic fibre fabrics that have long-term environmental burdens. The use of r-PET minimizes environmental waste from plastic bottles, while the mechanical properties of the non-wovens are further improved, ensuring that sustainability does not mean compromising on product life. It is finally suggested that green materials in demand from the construction sector may include insulation that, as shown, is environmentally favourable, does not incur additional financial costs, and at the same time contributes to solving the problem of the increasing heap of textile waste.

6.5.4 Case Study 4: Building Insulation Using Pineapple Fibre

The study on pineapple leaf fibre (PALF), an agro-residue collected from pineapple leaves, as an eco-friendly alternative to non-biodegradable composite insulation materials for sound and thermal insulation is reported in this study. The PALF was prepared by matt-raspador method, and the physical, morphological, and thermal properties are analyzed. The investigation aimed at the development of 100% PALF and PALF/low-melting point polyester (PET) as needle-punched non-wovens to improve performance (21).

The non-woven materials, featured for their porous structure, are capable of absorbing sound by causing the air molecules to vibrate within the cells and thereby lose energy. Sound absorption of PALF NWs is statistically evaluated for different frequency values. Although the 100% PALF non-wovens exhibited poor absorption at low frequency (100–400 Hz), they proved to be good absorbing materials at medium frequency (500–2,000 Hz). The absorption of sound at higher frequencies dropped somewhat, although it is still effective.

The higher GSM of non-wovens has better sound absorption performance, which is induced by their thicker structure that causes more energy to be dissipated while sound waves are passing through the material. The addition of PALF into the low-melting PET remarkably improved both the thermal and acoustic insulation abilities of non-wovens. The PET composite caused densification of the material with smaller pore size and higher flow resistance, which means that sound absorption, particularly at high frequencies, is improved. These enhancements are consistent with previous findings for the acoustics of fibrous structures.

The study found that 100% pineapple leaf fibre (PALF) non-woven, with a thickness of 1.03 mm, has a lower NRC of 0.29 compared to commercial glass wool. Blending 10% low-melt PET with PALF slightly improved the NRC to 0.35, likely due to increased compactness from PET bonding with the fibres. However, when 20% PET is added, the NRC dropped to 0.32. This decrease is attributed to higher PET content, which increased sound reflection, airflow resistance, and reduced porosity, leading to lower friction and sound energy loss. Overall, PALF non-wovens, especially when blended with PET, showed comparable sound insulation to commercial products. PALF's renewable nature and availability make it a promising, eco-friendly alternative for acoustic and thermal insulation, suitable for medium- and high-frequency sound absorption in building materials.

6.6 PACKAGING

Conventional packaging materials like plastic and Styrofoam are non-biodegradable and play a major role in environmental pollution. There is a growing need for sustainable packaging solutions that can replace these materials. Textile-based packaging, especially from recycled natural fibres like cotton and jute, offers a biodegradable and compostable alternative, addressing both waste management and environmental conservation.

6.6.1 Properties Required for Packaging

Generally, a packaging material is said to be effective if it carries properties such as durability, flexibility, and shock absorption to protect products during transit. In line with this, textile fibres possess all these qualities, which makes them suitable for packaging applications. Textile fibres can be processed into different forms, such as woven fabrics, non-woven mats, and composites, offering unique properties in applications.

6.6.2 Use of Textile Waste in Packaging

Nowadays, packaging materials made of recycled textile waste have gained significant interest. Unlike traditional packaging, which requires more energy for production processes, reusing textile waste can help in conserving natural resources and reducing carbon footprints. Textile waste can be tailored to create different forms of packaging, such as protective wraps, reusable bags, and cushioning materials, making it adaptable to diverse needs.

6.6.3 Pros and Cons of Textile Waste in Packaging

Pros: Textile-based packaging materials, especially from natural fibres, decompose faster than plastics. Recycling textiles can cut production costs by up to 30% by reusing existing materials instead of extracting new resources. Textile packaging can be reused multiple times, which is especially beneficial for industries like e-commerce, and can also replace single-use plastics. Textiles allow for easy customization, enabling companies to design unique, branded packaging.

Cons: The inconsistency in textile waste quality can affect the performance of packaging materials. Some textiles may have dyes or chemicals that need to be removed before repurposing, posing an additional processing challenge.

6.6.4 Applications of Textile Waste in Packaging

Textile waste is being used innovatively across various packaging solutions such as stuffing/cushioning, eco-friendly bags/pouches, corrugated boxes, etc. Textile waste can be shredded to create soft materials that protect fragile items during shipping, suitable for industries like electronics and ceramics. Recycled fabrics are used for making bags for retail goods, gifts, or promotional items, offering a sustainable alternative to plastic bags. Textile waste can be combined with cellulose materials to produce durable and biodegradable cardboard, used for food packaging and other applications.

6.6.5 Examples and Case Studies

According to the Ellen MacArthur Foundation, adopting reusable packaging solutions like those made from recycled textiles could eliminate about 20% of the plastic waste generated in the packaging sector annually. Studies indicate that turning textile waste into packaging materials could also help reduce landfill usage and contribute to environmental sustainability. Several companies are actively exploring and adopting textile waste for packaging:

- **Fruit leather (Netherlands):** Converts discarded fruit skins into leather-like fabrics used in packaging.
- **Repack (UK):** Creates reusable shipping bags from recycled textiles, reducing waste by enabling return, cleaning, and reuse up to 20 times.

- **India's experiments:** Trials with using jute and cotton waste to create biodegradable packaging materials have shown promising results, with potential to reduce landfill waste by up to 60%.

Textiles play a crucial role in the packaging industry, offering a wide range of functional and aesthetic benefits that enhance product protection and presentation. With growing environmental concerns and shifting consumer preferences, textile-based packaging solutions are gaining recognition for their versatility and functionality, while remaining eco-friendly. Textile packaging materials possess several essential properties that enhance their effectiveness, such as durability, customizability, and aesthetic appeal. With all these properties in mind, these packaging materials are employed across many sectors, which include food and beverages, retail and fashion, industrial applications, medical packaging, and so on. The properties of packaging materials can be enhanced with chemical finishes, which are determined by the specific end-use applications and requirements of the packaging materials. As environmental concerns become increasingly prominent, the textile packaging industry is embracing sustainable practices. Key trends include the use of biodegradable materials, eco-friendly finishes, reusability, and recycling initiatives.

6.6.5.1 Case Study 1: Corrugated Structure for Packaging Using Recycled Fibres

Thermally insulated packaging is essential for transporting temperature-sensitive products, including pharmaceuticals, fresh foods, flowers, and gourmet meals. This packaging is designed to regulate internal temperature and prevent heat transfer, preserving the product's quality during its journey. Corrugated structures are used in such packaging because of their structural strength, which safeguards products against shocks and compressive forces during transit. Corrugated board, recognized for its strength and cushioning, is widely used in packaging due to its ability to protect products during transport. The board's "flutes"—wave-like structures that come in varying sizes (from large A-flute to small F-flute)—provide strength and flexibility.

Larger flutes offer increased cushioning and durability, while smaller flutes enhance printability and flexibility, making it possible to customize boards for specific needs. The corrugated structure's design allows it to absorb and dampen vibrations, which is particularly beneficial during road transport, where vibrations pose a high risk of damaging packaged products. However, the primary disadvantage of conventional corrugated cardboard is its susceptibility to moisture, which weakens its structure and compromises its protective function. Leveraging recycled fibres to create these structures offers an eco-friendly solution for the packaging industry while supporting sustainability goals.

This study explores the design and development of a corrugated structure using recycled fibres from garment waste and recycled polyester (see Figure 6.15). The recycled fibres are converted into a non-woven felt using the processes of needle punching and thermal bonding. Low-melt polyester (LMF) fibres with a cut length of 32 mm are chosen for their bonding properties, which allow them to form a stable non-woven structure when combined with other fibres. To improve stiffness and

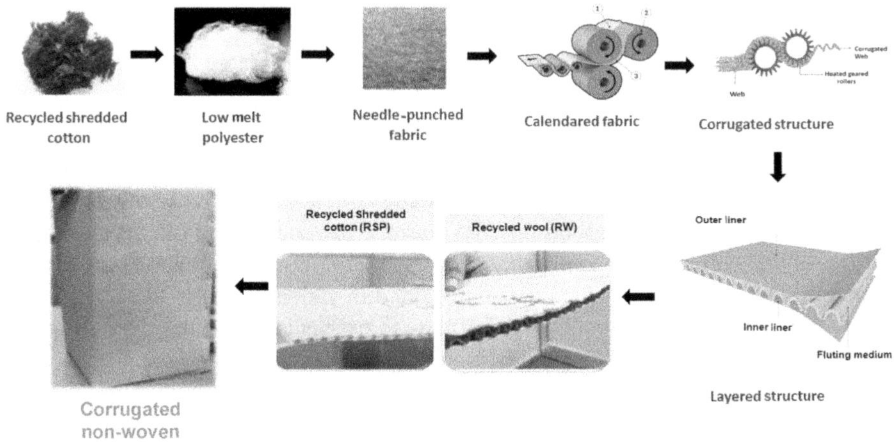

FIGURE 6.15 Development of corrugated structure using recycled fibres.

ensure stability, the needle-punched non-woven felt is hot calendared and further formed into a corrugated structure to create a flexible and lightweight material with excellent packaging strength and thermal insulation properties. The corrugation is achieved by applying pressure through grooved gears in an embossing machine, creating air pockets within the material. These air pockets provide insulation by minimizing heat transfer, thus preserving the internal temperature of the packaging. The thermally bonded non-woven (called thermal wadding) is then layered with the needle-punched corrugated structure in various configurations.

Conventional packaging materials commonly include plastics, paper, cardboard, glass, metals, and Styrofoam, offering various advantages in cost and functionality. However, their widespread use contributes to environmental pollution and waste. This invention addresses the negative environmental impact caused by non-recyclable and non-eco-friendly insulation materials commonly used in packaging industries and building insulation applications. The recycled textile material is sustainable and can be converted into box packages, providing temperature-controlled shipping solutions for various items, such as food products and pharmaceuticals.

6.6.5.2 Case Study 2: Packaging Using Banana Non-woven Fabric

Banana fibre has emerged as a promising natural substitute for synthetic materials as a result of the growing demand for biodegradable packaging. This case study investigates the utilization of the needle-punching technique for making non-woven packaging material from banana fibre (22). Because of its inherent UV protection, resistance to weather, and capacity to retain high moisture levels, banana fibre is advantageous for packaging and can help maintain the freshness of fruits and vegetables. Since India is the world's largest producer of bananas, the substantial amount of pseudostem waste produced during harvesting can be converted into useful packaging materials, lowering the pollution that traditional packaging materials cause to the environment.

Nowadays, synthetic materials like polypropylene, polyethylene, and thermoform are used to wrap a lot of fruits and vegetables. These materials are widely used, but they are not UV-protected and are not biodegradable, which raises additional environmental issues. Contrarily, because of their strength, light weight, affordable processing, and natural breakdown, natural fibres—like banana fibre—offer a more environmentally friendly substitute. Banana fibre, however, can be stiff and tightly bound, which makes it difficult to use. The study softened the fibres using enzymatic treatments in order to get around this. To improve the texture and quality of banana fibres and make them more appropriate for use in textiles, specific enzymes such as hemicellulose, pectinase, and cellulase are added in varied amounts.

By employing needle-punching technology to extract banana fibre from the pseudostem of the banana plant, the study is able to successfully create non-woven felt. Its efficacy as a packaging material is assessed by testing a variety of fruits and vegetables such as tomatoes, cucumbers, bananas, carrots, betel leaves, green chillies, and grapes to see if it could prolong their shelf life. Results showed that, in comparison to conventional packaging techniques, the majority of produce wrapped in banana fibre-based packaging stayed fresh for an extra one to five days.

On the other hand, bananas only displayed a two-day extension in freshness. The study also showed that applying effective microorganisms (EM) to banana fibres greatly increased their tensile strength. SEM, or scanning electron microscopy, verified that this procedure successfully eliminated wax and contaminants without seriously damaging the structure. Additional improvements were found to be promising ways to improve the banana fibre packaging's functionality and commercial appeal, including lowering the felt's thickness and adding antimicrobial coatings.

6.6.5.3 Case Study 3: Packaging Using Cotton Gin Trash and Polyvinyl Alcohol

This case study investigated the use of cotton gin trash (CGT) for manufacturing bioplastic for packaging applications. CGT is found abundantly in the cotton industry as a by-product often sent to landfill. The development of bio-plastic using CGT not only reduces landfill waste but also can be an economical and eco-friendly packaging material by replacing single-use plastics (23). The bioplastics are developed using CGT in conjunction with polyvinyl alcohol (PVA) to study the overall performance, flexibility, and water resistance. Ethyl cellulose (EC) is added as an adhesive to enhance the material's durability and water resistance.

PVA is dissolved at different concentrations (2–15%) with respect to total weight. Fine CGT powder and pre-dissolved PVA are mixed in a ratio of 50:50 for the fabrication process. The mixture is heated and stirred for proper dispersion and then poured onto polystyrene dishes to dry. According to the study, films plasticized with 10% DEG, 15% glycerol, or 15% urea are completely flexible and could fold more than 200 times without breaking. The water resistance of the films was significantly enhanced by the EC coating, with coated films demonstrating a minimum contact angle of 65°, effectively resisting water penetration for up to 10 min. Furthermore, the CGT/PVA composites exhibited superior UV protection, with a UV protection factor (UPF) rating between 219 and 229, indicating their suitability for outdoor applications.

The economic analysis estimated the production cost of these bio-plastic films to be approximately AUD 202–208 per kilogram, with the expenses primarily driven by the cost of chemicals. However, bulk purchasing could reduce costs further, making this approach viable for large-scale production. Overall, the study highlights the potential of CGT/PVA composites as a sustainable packaging solution, showcasing their flexibility, water resistance, and cost-effectiveness. By addressing key challenges associated with flexibility and water resistance, this research paves the way for practical applications of bioplastics in flexible packaging, contributing to efforts to mitigate plastic pollution and promote a circular economy within the textile and packaging industries. Future research will focus on scaling up production methods, particularly through extrusion technology, to enhance compatibility for industrial applications.

6.6.5.4 Case Study 4: Pulp-Moulded Packaging from Empty Fruit Bunch Fibre

The environmental challenges posed by single-use plastic packaging, including its extensive production, limited usage period, and difficult end-of-life management, have spurred the search for more sustainable alternatives. This study focuses on the potential of waste oil palm empty fruit bunch (EFB) fibres as a fibrous pulp for creating three-dimensional pulp-moulded products (24). By integrating compostable barrier treatments, these products can provide a fully bio compostable packaging solution, addressing the urgent need to mitigate plastic waste and promote environmental sustainability. The research highlights the feasibility of utilizing agricultural waste to develop eco-friendly packaging materials, effectively reducing reliance on non-biodegradable plastics.

The study involved converting waste EFB fibres into pulp using atmospheric refining, a mechanical process that eliminates the need for harmful chemicals. This fibrous pulp is subsequently used to produce pulp-moulded trays suitable for food packaging via a vacuum-forming method. Through optimized mould design, the study improved the efficiency of vacuum suction, enhancing fibre distribution and ultimately resulting in higher-quality trays. These trays are designed to hold various food items, including fruits and vegetables, demonstrating their practical application in the food industry. The physical properties of the trays are thoroughly evaluated, showing that the optimized design significantly enhanced the quality and fibre distribution, ensuring they are both recyclable and compostable.

As public and governmental concerns regarding plastic pollution rise, there is a growing demand for sustainable packaging alternatives. The study aligns with this trend, as over 60 countries have enacted regulations to limit plastic usage, prompting innovation in environmentally friendly packaging solutions. Traditional pulp-moulded products (PMPs) have commonly been derived from wood fibres, often facing contamination issues when recycled paper is used for food contact applications. In contrast, the study emphasizes the advantages of using EFB fibres—abundant by-products of the oil palm industry typically discarded or incinerated. The conversion of these fibres into pulp through atmospheric refining not only provides a high-quality fibre source for PMPs but also circumvents chemical contamination, enhancing the sustainability of the packaging. The potential applications for EFB fibre-based

PMPs extend beyond food containers to include clinical waste products and packaging for electronic goods. While this research demonstrates the viability of producing high-quality pulp-moulded trays from EFB fibres, further studies are recommended to align these products with specific food types and assess their performance in real-world packaging scenarios. Overall, this study shows how agricultural waste can be effectively utilized as a sustainable substitute for plastic packaging, encouraging global efforts to reduce plastic waste and promote the circular economy.

6.6.5.5 Case Study 5: Green Packaging from Isolation of Textile Waste Cellulose Nanofibrillated Fibre

The innovative production of biodegradable composites using cellulose nanofibrillated fibre (CNF) isolated from textile waste, particularly cotton-based fabrics, is emphasized in the case study on using textile waste as packaging material. This study tackles the urgent problem of plastic pollution brought on by synthetic polymers that are frequently used in packaging and greatly increase landfill and ocean waste. The approach used in this research involves separating CNF from waste textiles and mixing it with chitin and polylactic acid (PLA) to produce a biodegradable, sustainable composite for packaging uses (25). The textile waste-derived cellulose fibres serve as reinforcement in the PLA-chitin blend, significantly enhancing the mechanical and thermal properties of the final material. This composite presents a viable alternative to traditional plastic packaging, offering environmental benefits by diverting textile waste from landfills and utilizing it in creating eco-friendly materials.

Cellulose fibres from textile waste are used as reinforcement in the PLA-chitin blend, greatly improving the finished product's mechanical and thermal qualities. By keeping textile waste out of landfills and using it to make environmentally friendly materials, this composite offers a practical substitute for conventional plastic packaging.

Cotton textiles can be successfully converted into cellulose nanofibres through the use of a combination of high-pressure homogenization and supercritical carbon dioxide extraction in the process of separating CNF from textile waste. The addition of these fibres to the PLA-chitin matrix produces a composite with enhanced tensile strength, elongation, and impact resistance, which qualifies it for use in packaging. By increasing the material's flexibility and durability, the reinforcement with textile waste CNF overcomes these drawbacks. Furthermore, the hydrophilic CNF and the hydrophobic PLA are compatible with one another thanks to the hydrophobic chitin, which ensures even dispersion and enhanced miscibility in the composite matrix. The results of the study demonstrated that PLA, chitin, and CNF exhibited better performance than PLA alone or PLA-chitin blends, indicating that this ternary blend is a more promising material for use in green packaging applications.

By providing a solution to the twin problems of textile and plastic waste, the use of textile waste for packaging material is in line with global sustainability goals. Reusing this waste stream to make biodegradable packaging materials helps lessen the environmental impact of both industries, as textile waste is one of the biggest contributors to municipal solid waste. In addition, the PLA-chitin-CNF composite's biodegradable properties guarantee that the substance decomposes organically over time, allaying worries about plastic's environmental persistence. This investigation highlights the potential for developing high-performance, environmentally friendly packaging materials that not only reduce plastic pollution but also provide a circular approach to managing textile waste through enhancing the properties of biopolymers with cellulose nanofibres derived from textile waste. This case study serves as an example of how interdisciplinary approaches can result in sustainable innovations that support environmental conservation initiatives and the development of green technologies.

6.7 AGRO TEXTILES

Permeable textiles made from natural or synthetic fibres, geotextiles obtain wide application in environmental, construction, and civil engineering applications. Enhancing soil performance, safeguarding structures, and controlling water flow are the main purposes of geotextiles. They are employed to strengthen structures, improve drainage, reduce erosion, and increase soil stability. Roads, embankments, retaining walls, and foundations can all be built with them. Geotextile-based erosion control mats aid in soil stabilization and encourage the growth of vegetation. Natural fibres like jute and coir, as well as synthetic fibres like polyester and polypropylene, are among the materials used to make geotextiles. Both producers and end users can gain from the affordable alternatives to conventional materials that geotextiles made from recycled textiles can provide.

6.7.1 Mechanism of Geotextiles

The vital functions of geotextiles and their mechanisms are as follows: separation, filtration, drainage, reinforcement, protection, and erosion control.

- **Separation:** When two different layers of soil or other materials are separated, geotextiles serve as a barrier. This keeps various soil layers from combining, which is essential for preserving the strength and integrity of structures. Geotextiles are used in roads, railroads, embankments, and foundations to keep soil layers apart and stop the subgrade from blending with the base material.
- **Filtration:** Water can flow through geotextiles, but fine soil particles cannot. This filtration feature aids in controlling water flow and guards against erosion and drainage system blockages.
- **Drainage:** Geotextiles provide an efficient pathway for water to move through the soil, facilitating drainage. They act as conduits, allowing water to be transmitted in the plane of the fabric (lateral drainage). This helps in managing excess water within soil structures, preventing water clogging and maintaining stability.
- **Reinforcement:** When used as a reinforcement layer, geotextiles distribute applied loads more evenly across the soil, improving tensile strength. The geotextile fabric interacts with the surrounding soil through friction or adhesion, enhancing its ability to resist tensile forces and limiting deformation.
- **Protection:** Geotextiles protect soil structures or membranes from damage by external forces. In landfill liners, tunnels, and waterproofing systems, geotextiles are used to protect geo-membranes from sharp objects or pressure, maintaining the integrity of the sealing layer.

- **Erosion Control:** Geotextiles help control erosion by stabilizing soil and reducing the displacement of soil particles due to water or wind forces. They slow down the flow of water across the soil surface, reducing its capacity to carry soil away. This prevents surface erosion and helps vegetation grow by stabilizing the soil.

6.7.2 Utilization of Textile Waste in Geotextiles

The fibrous structure of textile waste makes it ideal for use as geotextiles. Textiles, especially from natural fibres like cotton and jute, and synthetic fibres like polyester, have high tensile strength, flexibility, and durability—qualities essential for geotextiles. When processed and woven appropriately, recycled textile fibres can effectively reinforce soil, control erosion, and improve soil stability.

6.7.3 Benefits of Using Textile Waste as Geotextiles

- **Cost-effective:** Reusing textile waste to create geotextiles lowers production costs and the demand for raw materials. Recycled textiles can be processed using less energy and resources than traditional geotextiles, which need energy-intensive manufacturing processes. Textile-based geotextiles are a financially appealing alternative since, according to a European Commission report, employing recycled materials can lower production costs by as much as 30%.
- **Environmental benefits:** The quantity of waste that ends up in landfills or incinerators is decreased when textile waste is recycled into geotextiles. Additionally, this reduces the carbon footprint linked to waste management and new material production. For instance, recycling one ton of cotton can cut greenhouse gas emissions by 2.1 tons and save up to 765 m^3 of water.
- **Effective soil and water management:** Geotextiles made from textile waste have been effectively applied to a number of applications, such as water filtration, road construction, and soil erosion control. For instance, jute- or coir-based natural fibre geotextiles can aid in riverbank and slope stabilization, stop soil erosion, and promote vegetation growth. Researchers in India have created geotextiles from recycled jute that have been applied in practical settings to stabilize embankments and preserve riverbanks.

6.7.4 Challenges and Future Prospects

Although there are many advantages to using textile waste to create geotextiles, there are also issues that must be resolved. Variability in textile waste's quality and characteristics is a significant issue since it may have an impact on the geotextiles' performance and consistency. Additionally, some textiles might include chemical treatments or dyes that, if improperly handled, could be hazardous to the environment. By making it possible to separate and purify textile fibres and guarantee that they fulfil quality requirements for geotextile applications, developments in textile recycling technologies are aiding in resolving these problems.

As more businesses and academic institutions look into innovative ways to use textile waste in civil engineering projects, the market for recycled geotextiles is expanding. For example, in Europe, initiatives have been taken under the programme called "Horizon 2020," mostly focusing on developing sustainable geotextiles from recycled materials, with the aim of reducing the environmental impact of construction projects.

6.7.4.1 Case Study 1: Geotextiles for Erosion Protection Using Textile Wastes

In a recent study, thick ropes designed to prevent slope erosion are made from textile waste. Non-woven textiles made from recycled fibres and leftover insulating material are used to make the ropes. Over the course of two vegetation seasons, the ropes' efficacy is assessed after they are erected on a steep slope in an abandoned gravel pit.

Two types of non-woven strips are used: Wool non-woven made from coarse wool fibres, primarily used for insulation. These fibres, unsuitable for high-quality textiles, are repurposed for slope protection. Recycled Fibres Non-woven made from a blend of polyester, cellulose, and wool fibres from post-consumer apparel waste (26). Various fibres are combined using a stitching method. These strips are twisted into ropes with a linear density of about 740–745 g/m and encased in a polypropylene sheath. The ropes are then organized into segments and secured with linking chains. The ropes are installed transversely on the slope, covered with topsoil, and monitored for fibre morphology and mechanical properties. The ropes provided immediate slope stabilization, facilitating soil particle settlement and integration with the native ground. They also prevented surface water erosion and improved water management, enhancing seed germination and plant growth.

The wool non-woven ropes biodegraded slowly in the humid soil, maintaining their protective function for two vegetation seasons. This allowed time for mature vegetation to establish and take over the protective role. Ropes made from recycled fibres showed minimal biodegradation over two years, offering long-term protection due to their resistance to biological and chemical degradation. This case study demonstrates the feasibility and benefits of using textile waste for erosion protection. The technology not only repurposes poor-quality materials and industrial waste but also provides an effective solution for slope stabilization. By converting textile waste into valuable geotextiles, the total amount of waste is minimized, contributing to sustainable waste management practices.

6.7.4.2 Case Study 2: Jute Non-woven for Seed Germination

This case study investigates the creative application of low-grade jute fibre as a soil substitute for seed germination in the form of needle-punched non-woven fabric (27). This fabric's bulk density affects root growth, thermal insulation, and the penetration of water and nutrients, making it essential for agricultural use. By altering the punch density while maintaining the same area density and needle penetration depth, different bulk densities are produced. To make a fibre web, jute reeds are processed through a variety of machines, softened, and conditioned. Jute scrim fabric is used to reinforce this web, and a needle loom is used to turn it into a non-woven fabric. The ideal conditions for seed germination are determined by testing various bulk densities. The best results are observed with a fabric of 500 g/m^2 area density and 0.116 g/cm^3 bulk density, achieved using 160 $punches/cm^2$ and a 13 mm depth of needle penetration.

In comparison to soil, the non-woven fabric demonstrated better moisture retention and temperature control, both of which are critical for plant growth in harsh conditions. According to the study, plants grown in non-woven fabric outgrew those grown in soil in terms of growth and survival rates. In particular, the ideal conditions are offered by fabric having a bulk density of 0.116 g/cm^3. The non-woven fabric absorbed water more slowly but retained it longer than soil, promoting uniform moisture distribution through the wicking effect of jute fibres. Farmers can easily handle the simple process of preparing the non-woven fabric. An important issue in agriculture is soil erosion, which the fabric stops. The non-woven fabric improves resource use efficiency by lowering water and fertilizer loss. Jute adds nutrients to the soil as it breaks down. Because the fabric is reusable, it is a sustainable choice. Transportation and handling are made easier by the fabric's lower weight when compared to soil.

From the study, it is clear that the use of needle-punched non-woven jute fabric for seed germination will be a viable alternative. This innovative approach can potentially revolutionize agriculture by providing a controlled environment for plant growth, reducing soil erosion, and enhancing resource efficiency.

6.7.4.3 Case Study 3: Geotextiles Using Textile Waste

This case study supports the "circular economy" concept, focusing on creating a closed-loop system in the Romanian textile industry by recycling polyacrylonitrile (PAN) knitted products and polyester (PES) woven patches. The primary goal is to produce non-woven geotextiles, which offer numerous advantages in various applications, including railway reinforcement, slope stabilization, erosion control, drainage, filtration, road paving, and crop coverings. The traditional linear model of "take-make-consume-dispose" in the textile industry results in significant waste, both post-consumer (collected from the population) and pre-consumer (from textile manufacturing). The circular textiles industry aims to reimagine this model, promoting the reuse and recycling of textile materials. This involves separating different types of materials before disposal and collecting technological waste separately based on raw material type.

Polypropylene (PP) fibres are often used in geotextiles due to their advantageous properties, such as lightweight, low moisture absorption, good mechanical resistance, and high chemical and corrosion resistance. The research aimed to produce non-woven synthetic geotextiles by recycling PAN knitted products and PES woven patches. The PAN fibres constituted over 75% of the recycled fibres used in production. The addition of PP fibres aimed to improve the properties of the non-woven geotextiles and reduce variability. Three fibrous blends are created using recovered fibres (PES and PAN) and first-use fibres (PP) in different ratios (28).

Recycled fibres tend to be physically damaged, shorter, and have a high variation in properties compared to first-use fibres. As a result, most experimental data variations are lower with the addition of more PP fibres, except for bulk density. Despite the physical limitations of recycled fibres, the study confirmed the feasibility of producing high-quality non-woven geotextiles by incorporating recycled fibres into fibrous blends. This approach reduces raw material costs and supports the circular economy concept.

The study showcases the feasibility of adopting the circular textiles industry model in Romania by establishing regional closed-loop systems. These systems would reuse all types of textile materials to manufacture non-woven geotextiles. Training for both the population and companies is essential to ensure the correct collection and separation of textile waste. Recycling polyacrylonitrile knitted products and polyester woven patches to produce non-woven geotextiles supports the circular economy by reducing waste, lowering raw material costs, and promoting sustainability. By developing a competitive and sustainable business model, the circular textiles sector has the potential to have a big impact in Romania. Achieving these objectives and optimizing the advantages of recycled textile materials in the creation of geotextiles depend heavily on proper implementation and training.

6.7.4.4 Case Study 4: Geotextiles from Reclaimed Fibres

Erosion-prone slopes are frequently the result of environmental damage brought on by earthworks close to mines, roads, and other man-made structures. This study explores the use of geotextiles derived from recycled fibres, particularly wool and polypropylene, as an environmentally friendly way to stabilize slopes and encourage the growth of vegetation. A major worldwide problem, textile waste encompasses a variety of post-industrial, pre-consumer, and post-consumer waste types. For the sustainability of the economy and the environment, recycling these materials is essential. These textiles are converted into loose fibres by the study's mechanical recycling techniques, and these fibres are then used to make non-woven geotextiles that improve soil qualities and slope greening.

The study, which is carried out at four distinct locations in southern Poland's Silesia region, concentrated on installing geotextiles made of recycled fibres to stabilize slopes with different substrata, including gravel and clay. As part of the experimental setup, meandering rope segments are prepared and secured to the slopes before being covered with topsoil. In order to examine floristic composition, species diversity, and vegetation growth, the study set up 46 plots. According to the research, some plant species did better in environments with a higher wool content, while others fared better in environments with polypropylene (29). In comparison to control plots, the presence of reclaimed geotextiles is linked to increased plant height and colonization rates, suggesting that they have a beneficial effect on slope greening. The evenness of species improved even though overall species diversity decreased, as indicated by the Shannon-Wiener diversity index, indicating that mechanical stabilization via rope use is crucial in fostering plant growth.

The study emphasizes the various ways that repurposed fibre geotextiles can improve soil quality and support vegetation. Due to their decomposition, wool fibres improved soil fertility by increasing the amount of nitrogen in the soil. On the other hand, polypropylene performed better at increasing species diversity, possibly as a result of its unique nutrient dynamics and moisture retention. By functioning as micro-dams, the creatively designed ropes composed of a blend of hydrophilic and hydrophobic fibres efficiently controlled water on slopes. This design promoted vegetation growth, particularly during drought conditions, by reducing runoff, retaining moisture, and guaranteeing uniform water distribution. The study suggests that a higher wool content may favour dominant plant species, lowering biodiversity even though soil stability and erosion reduction have improved overall. Reclaimed fibre geotextiles have a lot of promise for stabilizing slopes and promoting quick vegetation growth, but more long-term study is required to completely comprehend their implications for habitat development and biodiversity preservation. Because agrotextiles offer a sustainable substitute for environmental restoration initiatives and support the circular economy in the textile sector, this study emphasizes the significance of investigating their use.

6.8 FILTRATION

By passing a liquid or gas mixture through a filter medium that lets the fluid flow while trapping the solid particles, filtration removes solid contaminants from the mixture. These contaminants can be eliminated using a variety of methods.

6.8.1 Types of Filtrations

Two broad categories can be used to classify filtration procedures involving textile waste, namely surface and depth filtration. Surface filtration traps particles on the surface of a porous material, like fabric. This technique works well for catching bigger particles and is frequently applied in industrial dust collection and HVAC systems. In depth filtration, particles are trapped inside a thick, porous medium. Oil filtration and wastewater treatment are two uses for this technique, which works well for capturing smaller particles.

6.8.2 Mechanism of Filtration

Figure 6.16 illustrates the various mechanisms by which a fibrous filter captures particles from any fluid stream. Fibrous filters are widely used in air/liquid filtration systems because of their ability to trap particles of different sizes through multiple physical and electrostatic interactions. The mechanism includes inertial impaction, interception, Brownian motion, electrostatic attraction, and gravitational settlement. Each of these plays a significant role in the overall filtration process.

1. **Inertial impaction:** Inertial impaction occurs when larger particles, due to their higher mass and inertia, are unable to follow the streamline (path of the airflow) as it bends around the fibres of the filter. Instead of moving with the air, these particles continue along their original path and collide with the fibres. This is particularly effective for capturing larger particles that are unable to change direction quickly.
2. **Interception:** Interception happens when a particle, while following the airflow streamlines, comes close enough to a fibre that it gets captured. Unlike inertial impaction, interception does not rely on the particle's inertia. Instead, the particle simply touches the fibre while moving along its path and gets stuck. This mechanism is effective for mid-sized particles that can follow the airflow but are still large enough to make contact with the fibres.

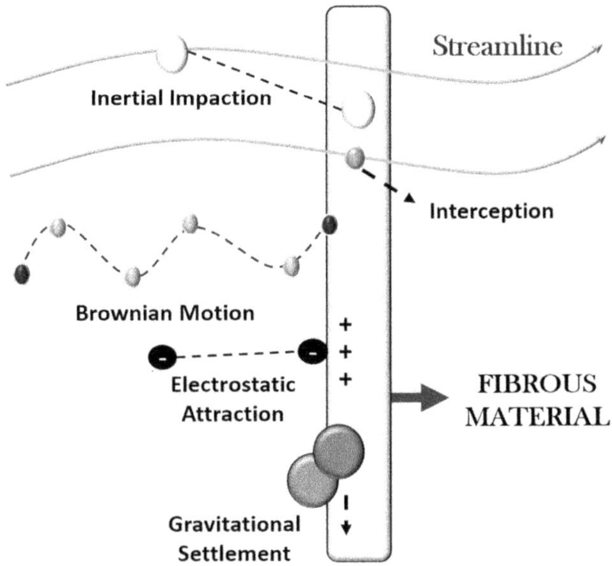

FIGURE 6.16 Mechanism of filtration process.

3. **Brownian motion:** The random, irregular movement of extremely small particles (less than 0.1 μm) caused by collisions with gas molecules is known as Brownian motion. These particles are more likely to come into contact with the filter's fibres and be captured because of their random motion. Because it increases the likelihood that ultrafine particles will be intercepted, Brownian motion is especially important for their removal. The tiny particles travelling in a zigzag pattern in the figure stand in for particles undergoing Brownian motion.
4. **Electrostatic attraction:** Particles are drawn together by electrostatic forces. The fibres in fibrous filters have the ability to carry an electric charge, which draws in particles with opposing charges. Furthermore, when neutral particles approach a charged fibre, they may become polarized, which results in electrostatic attraction. By capturing particles that might otherwise pass through, this mechanism increases the filter's efficiency. The picture illustrates how charged particles and fibres interact, emphasizing the operation of electrostatic forces.
5. **Gravitational settlement:** The process by which heavier and larger particles adhere to the filter fibres as a result of gravity is known as gravitational settlement. When particles are heavier and denser, this mechanism works better, causing them to drift downwards and land on the filter instead of following the airflow. Gravitational settlement aids in the capture of larger particles but is not as important as the other mechanisms. The picture depicts a big particle falling onto the fibre as a result of gravity.

Fibrous filters efficiently capture particles of different sizes by combining physical and electrostatic interactions. Fibrous filters can capture a wide range of particles, from larger debris to ultrafine pollutants, thanks to the cooperation of these mechanisms. Designing and improving air filtration systems to enhance air quality in a variety of settings, from homes to businesses, requires an understanding of these mechanisms.

6.8.3 Use of Fibrous Structure as a Filter

Textile fibres can be transformed into useful filtration media using a variety of techniques. Threads are interlaced during the weaving process to produce a fabric that can be used as a filter. This process can create effective and long-lasting filters for a variety of uses. Textile fibres are compressed during felting to create a thick, mat-like substance. Because felt textile waste is thick and porous, it can be used for depth filtration. Fibres are mechanically entangled by needle punching to produce a non-woven fabric. Filtration media with variable thickness and porosity that are appropriate for both surface and depth filtration can be produced effectively using this process. The goal of ongoing research and development is to improve the functionality and adaptability of filtration media. The use of nanotechnology is one area of progress. To increase filtration performance and efficiency, textile-derived nanofibres can be added to filtration media. Because of their high surface area-to-volume ratio, nanofibres are capable of effectively capturing even the tiniest particles.

6.8.4 Application of Textile Fibres in Filtration

Textile fibres are used in a wide range of industries and applications, such as medical, liquid, and air filtration. Filters for industrial dust collection systems, automobile air filters, and HVAC systems can be made from textile fibres. By capturing dust, pollen, and other airborne particles, these filters contribute to better air quality. Additionally, they can be used in liquid filtration processes like food and beverage processing, oil filtration, and wastewater treatment. These filters efficiently eliminate impurities and pollutants from liquids, guaranteeing clean and safe water or fluids. Fibres can be recycled into surgical gowns, face masks, and wound dressings for use in the medical field. These goods support health and safety regulations by offering crucial filtration and protection.

6.8.5 Use of Textile Waste as a Filter Media

Textile waste, which mostly consists of used clothes and textiles, can be recycled into a useful filter for water and air purification systems. This not only aids in waste management but also offers a sustainable and affordable answer to filtration requirements. Waste from textiles, particularly cotton, polyester, and other synthetic materials, has properties that make it a good filter. It is possible to alter the fibrous structure of these textiles to improve their filtration capabilities. Additionally, textiles have a high porosity, which helps to absorb pollutants and particulates. Cotton textiles, for instance, have been found to be able to filter particles as small as 0.3 μm, which is comparable to many commercial filters. Waste polyester can also be processed and woven into mesh constructions.

6.8.6 Advantages of Using Textile Waste as Filters

- **Cost-effective:** By using textile waste as a filter, less raw material is needed, which lowers costs. Conventional filter media production frequently calls for the use of costly synthetic polymers and intricate manufacturing procedures. Manufacturers can reduce production costs by using recycled textiles, which will make filters more affordable, particularly in areas with low incomes.
- **Environmental benefits:** The quantity of waste that ends up in landfills and incinerators is decreased by reusing textile waste. This helps to reduce greenhouse gas emissions linked to waste disposal in addition to reducing the environmental impact of textile waste. One ton of recycled cotton, for example, can save up to 20,000 L of water, which is a resource-intensive aspect of cotton production.
- **Versatility and customization:** Waste from textiles can be processed to meet different filtration needs. For instance, filters for water treatment, air purification, or even industrial gas scrubbing can be made using various fabric structures. Textiles are versatile for various filtration designs due to their inherent qualities, which include tensile strength, flexibility, and absorbency. For instance, waste denim has been successfully used by researchers to make filters that can remove oil from water with an efficiency of more than 90%.

6.8.7 Challenges and Future Prospects

The use of textile waste as filters has drawbacks despite the possible advantages. One problem is that textile waste varies in composition and quality, which can have an impact on the filters' performance and consistency. Furthermore, some fabric treatments and dyes may introduce contaminants that must be eliminated before the textile can be used again for filtration. However, processing and purifying textile waste for particular uses is becoming simpler thanks to developments in textile recycling technologies. Various initiatives worldwide are working to scale up the use of recycled textiles in filtration systems. For example, in India, researchers have developed air filters from recycled cotton that can capture PM2.5 particles with an efficiency of over 80%. In Europe, projects are focusing on turning textile waste into filter media for wastewater treatment, offering a sustainable solution for managing both textile waste and water pollution.

6.8.7.1 Case Study 1: Jute-Based Filters for Industrial Effluent Filtration

Prefilters are placed before the tertiary effluent process, namely reverse osmosis. Reverse osmosis is carried out to get pristine pure water from the influent. The success of the tertiary process in terms of rejection lies in feeding influent free from turbidity. The turbidity of the influent, if maintained at <1 NTU (Nephlometric Turbidity

Units), yields better performance in reverse osmosis plants. Currently, prefilters are available as string wound and melt blown cartridges. The fibres used in the prefilters are polypropylene and glass fibres depending on the type of effluent processed. In the present scenario, synthetic polypropylene roving in wound/cross wound manner is used in textile industries for effluent treatment. The filtration efficiency of the polypropylene roving is not sufficient. The biggest limitation for cartridge filters is their inability to treat waste streams with solid loadings greater than 1 or 2 mg/L for a standard length 10 inch. Also, they are not recyclable, leading to environmental pollution.

This study examines the potential of jute fibre-based non-wovens for effluent prefiltration. Jute non-woven fabrics are calendared and developed into cartridge filters. Also, a string wound filter is developed using the yarn spun from jute fibre (see Figure 6.17). These prefilters effectively capture large particles, reducing the load on advanced filtration systems. The filtration efficiency of the developed cartridge filters from jute non-woven fabric is found to be higher than that of the commercial filter. In general, jute fibrous non-woven-based prefilter would be a good alternate for present polypropylene-based filters. The findings highlight jute-based filters as sustainable alternatives to synthetic options, promoting eco-friendly practices in industrial operations.

6.8.7.2 Case Study 2: Oil-Water Separation Using Raw Cotton Fabric

Oil-water separation is a critical process in environmental management, especially in industries dealing with oil spills or wastewater treatment. Traditional filtration membranes often face challenges in terms of durability, cost, and performance under harsh conditions. This case study focuses on the development of a cotton fabric treated with a hybrid coating of mesoporous silica and epoxy resin to create a low-cost, durable, and highly efficient membrane for oil-water separation applications (30).

The cotton fabric is pretreated with 5% sodium hydroxide solution to remove sizing materials and then washed, followed by drying at 80°C. A hardener and coupling

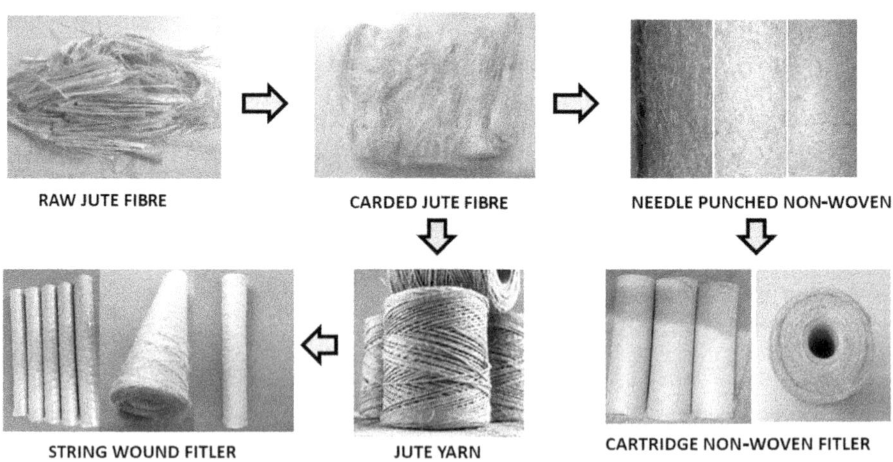

FIGURE 6.17 String wound and cartridge type prefilter for effluent wastewater treatment.

agent are added to a mixture of 10% epoxy and 1% SBA-15 for coating the fabric. To ensure a uniform coating, the fabric is submerged in the solution and then padded at a pressure of 2 kg/cm^2. To completely crosslink the epoxy, the treated fabrics are subsequently dried at 80°C and cured for five minutes at 180°C.

The coated cotton fabric's high hydrophobic and superoleophilic properties are confirmed by its low surface free energy (SFE) of 13.9 mN/m and water contact angle (WCA) of 146°. When tested with a diesel-water mixture, the fabric demonstrated a separation efficiency of 95.8% and a flux rate of 6,800 L/m^2 h, significantly outperforming untreated cotton and single-coated fabrics. It also showed an excellent UPF of 29, indicating its effectiveness in shielding against harmful UV radiation. Long-term durability is also ensured by the fabric's strong resistance to washing, mechanical abrasion, and pH changes.

Mesoporous silica and epoxy resin worked in concert to produce a strong hybrid coating that improved the performance of the cotton fabric without sacrificing its flexibility. The fabric's superior hydrophobic and oleophilic qualities are facilitated by the rough, microvillus-aggregated surface morphology seen in SEM images. An efficient solution for oil-water separation applications is the silica/epoxy-coated cotton fabric (SE/cotton), which provides a long-lasting, reasonably priced membrane with superior hydrophobicity, oleophilicity, and UV protection. It is a strong contender for practical industrial uses due to its long-lasting qualities and high separation efficiency, especially in environmental filtration systems. This finding demonstrates the potential of mesoporous silica and epoxy-based coatings in advancing textile materials for specialized filtration applications.

6.8.7.3 Case Study 3: Filter Media Using Recycled Coconut Fibre and Rice Husk

Water quality is seriously threatened by rising contamination as urbanization and industrialization increase, making efficient wastewater treatment techniques essential to maintaining and enhancing water supplies. This study explores the use of rice husk and coconut fibre, two recycled agricultural wastes, as filter materials for wastewater treatment. Evaluating the contaminants found in wastewater and determining how well these materials remove pollutants are the main goals. In addition to managing wastewater, this strategy makes use of the wealth of agricultural waste materials, especially in nations like Bangladesh where they are easily accessible.

Rice husk and coconut fibre are chosen due to their accessibility and qualities. Rice husk's chemical makeup and granular structure make it an efficient adsorbent, while coconut fibre's high surface area and microporous structure can prevent fungal growth. The filter setup used in the study is made up of layers of sand and stone chips, with the middle layer made up of rice husk and coconut fibre. To determine the efficacy of the filter media, wastewater samples are gathered from residential areas, and inflow and outflow water quality parameters—such as turbidity, total suspended solids, and coliform counts—are examined. Both rice husk and coconut fibre considerably enhanced water quality, according to the results, proving that they are viable low-cost, environmentally friendly substitutes for conventional filtration techniques (31).

The results showed that rice husk and coconut fibre successfully decreased turbidity, suspended solids, and coliform counts, thus satisfying Bangladeshi water quality standards. Their complementary functions in the filtration process are demonstrated by the superior removal of faecal coliform by coconut fibre and the superior removal of total coliform by rice husk. In addition to providing a more affordable option than traditional techniques like activated carbon, this study highlights the sustainable use of agricultural waste in wastewater treatment, repurposing agricultural residues to address waste management concerns. In conclusion, treating wastewater with rice husk and coconut fibre as filter media shows a lot of promise and represents a major advancement in sustainable water management techniques. This case study reinforces the importance of exploring innovative, eco-friendly solutions for wastewater treatment, promoting the utilization of locally available agricultural waste to enhance water quality and create a healthier environment for future generations.

6.9 SUMMARY

In the fashion industry, textile waste is a major problem, but it also offers a lot of potential for value addition and a variety of uses. Textile waste must be managed effectively, which calls for both systemic and individual solutions. By embracing sustainable clothing practices, such as buying from eco-friendly brands, recycling, and upcycling used clothing, consumers play a critical role in this process. The total environmental impact of the fashion industry can be considerably decreased by these individual efforts.

The fashion industry needs to invest in cutting-edge recycling technologies, embrace the circular economy, and integrate sustainable production methods in order to reduce textile waste on a systemic level. In addition to reducing environmental harm, these steps create new opportunities for revenue streams and business models. Value addition can turn waste into resources by turning waste materials into new fibres, fabrics, or other products. This demonstrates a company's dedication to sustainability and meets the growing demand for eco-friendly products.

Beyond the fashion industry, recycled textiles find use in the home décor, automotive, and construction sectors, among other industries. New apparel, accessories, and shoes that are notable for their sustainability can be made from recycled textiles. These materials can be made into upholstery, curtains, and rugs for home décor, giving customers environmentally friendly interior design choices. Recycled textiles can be used as reinforcement, insulation, and soundproofing materials in the construction and automotive sectors, reducing waste and conserving natural resources.

The shift to a circular textile economy is greatly aided by governmental and regulatory organizations. The fashion industry can be pushed towards more ethical and ecologically friendly practices by policies that promote the use of recycled materials, encourage sustainable production methods, and control waste management. Additionally, funding for the study and creation of cutting-edge recycling technologies can hasten this shift and promote a more environmentally friendly sector.

In order to address textile waste, awareness and education are crucial. Customers are better equipped to make decisions regarding the acquisition and disposal of clothing when they are informed about the negative effects of the fashion industry on the environment as well as the advantages of recycling and upcycling. Furthermore, more sustainable supply chains, less waste, and a greater use of recycled materials can result from teaching designers, manufacturers, and retailers about sustainable practices and the importance of textile waste.

In the final analysis, the fashion industry faces many obstacles due to textile waste, but there are also many chances for long-term, sustainable growth. The textile industry can lessen its environmental impact and transition to a circular economy by adopting sustainable practices, making investments in recycling technologies, and raising consumer awareness. In order to maximize the value addition and application of textile waste and create a more environmentally friendly and sustainable fashion industry, cooperation amongst stakeholders, including companies, governments, and consumers, is essential.

REFERENCES

1. Mempouo B, Cooper E, Riffat SB. Novel window technologies and the code for sustainable homes in the UK. *International Journal of Low-Carbon Technologies.* 2010 Dec 1;5(4):167–74.
2. Global Alliance for Buildings and Construction. *Global Status Report for Buildings and Construction.* Nairobi, Kenya: United Nations Environment Programme; 2020.
3. Bentoumi L. Impact of thermal insulation on energy consumption in buildings. *Journal of Thermal Engineering.* 2024;10(4):924–35.
4. Patnaik A, Mvubu M, Muniyasamy S, Botha A, Anandjiwala RD. Thermal and sound insulation materials from waste wool and recycled polyester fibers and their biodegradation studies. *Energy and Buildings.* 2015 Apr;92:161–9.
5. Muthu Kumar N, Thilagavathi G, Karthikka M. Development of recycled PET/comber noil non-wovens for thermal insulation application. *Journal of Natural Fibers.* 2022 Sep 2;19(9):3233–40.
6. Muthukumar N, Thilagavathi G, Kiruba T. Effect of bonding techniques on thermal insulation and biodegradation of banana fiber/recycled polyester nonwovens. *Journal of Reinforced Plastics and Composites.* 2024 Mar;43(5–6):293–300.
7. Thenmozhi R, Thilagavathi G. Effect of kapok/polypropylene needle punched nonwoven and chicken feather fiber wadding on thermal insulation applications. *The Journal of The Textile Institute.* 2024 Nov;115(11):2059–65.
8. Kumar N M, Thilagavathi G, Periasamy S, Vinoth V. Development of needle punched nonwovens from natural fiber waste for thermal insulation application. *Journal of Natural Fibers.* 2022 Oct 28;19(14):9580–8.
9. Muralidharan V, Thilagavathi G. Investigation of sound absorption properties of hybrid multilayer density gradient coir structure. In: *Proceedings of the First International Conference on Combinatorial and Optimization*, ICCAP 2021, December 7–8 2021. Chennai, India: EAI; 2021 [cited 2024 Dec 25]. Available from: http://eudl.eu/doi/10.4108/eai.7-12-2021.2314624
10. Thilagavathi G, Muralikrishnan A, Muthukumar N, Neela Krishnan S. Investigation on sound absorption characteristics of nonwoven coir mats. In: Majumdar A, Gupta D, Gupta S, editors. *Functional Textiles and Clothing [Internet].* Singapore: Springer Singapore; 2019 [cited 2024 Dec 25]. p. 37–45. Available from: http://link.springer.com/10.1007/978-981-13-7721-1_4
11. Thilagavathi G, Neela Krishnan S, Muthukumar N, Krishnan S. Investigations on sound absorption properties of luffa fibrous mats. *Journal of Natural Fibers.* 2018 May 4;15(3):445–51.

12. Thilagavathi G, Pradeep E, Kannaian T, Sasikala L. Development of natural fiber nonwovens for application as car interiors for noise control. *Journal of Industrial Textiles.* 2010 Jan;39(3):267–78.
13. Viju S, Thilagavathi G, Keertana N. Oil sorption properties of cotton comber noil/recycled polyester blended needle punched nonwoven fabrics: An optimization study. *Journal of Industrial Textiles.* 2022 Jun;52:15280837221113090.
14. Viju S, Rengasamy RS, Thilagavathi G, Singh CJ, Mohamed HAK. Sustainable development of needle punched nonwoven fabrics from silk worm cocoon waste for oil spill removal. *Journal of Natural Fibers.* 2022 Nov 2;19(11):4082–92.
15. Subramoniapillai V, Thilagavathi G. Utilization of recycled polyester nonwovens as sorbent for oil spill cleanups. *Research Journal of Textile and Apparel.* 2021 May 31;25(2):158–69.
16. Thilagavathi G, Praba Karan C, Das D. Oil sorption and retention capacities of thermally-bonded hybrid nonwovens prepared from cotton, kapok, milkweed and polypropylene fibers. *Journal of Environmental Management.* 2018 Aug;219:340–9.
17. Thilagavathi G, Praba Karan C. Investigations on oil sorption capacity of nettle fibrous assembly and 100% nettle and nettle/kapok blended needle-punched nonwovens. *Journal of Industrial Textiles.* 2019 Oct;49(4):415–30.
18. Teklehaimanot M, Hailay H, Tesfaye T. Manufacturing of ecofriendly bricks using microdust cotton waste. *Journal of Engineering.* 2021 May 6;2021:1–10.
19. Mohan S, Thilagavathi G, Rajkhowa R. Composite panels utilizing microdust and coir pith for eco-friendly construction solutions. *The Journal of The Textile Institute.* 2024 Mar 4;116(2):1–8.
20. Muthukumar N, Thilagavathi G. Development of sustainable thermal and sound insulation materials from textile fiber wastes. In: Arun Ramnath R, Sanjay MR, Sergey M, Gorbatyuk SF, Suchart S, editors. *Biocomposites—Bio-Based Fibers and Polymers from Renewable Resources* [Internet]. Oxford, UK: Elsevier; 2024 [cited 2024 Dec 25]. p. 247–256. Available from: https://linkinghub.elsevier.com/retrieve/pii/B9780323972826000116
21. Thilagavathi G, Muthukumar N, Neela Krishnan S, Senthilram T. Development and characterization of pineapple fibre nonwovens for thermal and sound insulation applications. *Journal of Natural Fibers.* 2020 Oct 2;17(10):1391–400.
22. Manickam P, Kandhavadivu P. Development of banana nonwoven fabric for eco-friendly packaging applications of rural agriculture sector. *Journal of Natural Fibers.* 2022 Aug 3;19(8):3158–70.
23. Haque ANMA, Naebe M. Flexible water-resistant semi-transparent cotton gin trash/poly (vinyl alcohol) bio-plastic for packaging application: Effect of plasticisers on physicochemical properties. *Journal of Cleaner Production.* 2021 Jun;303:126983.
24. Liu Q, Loxton C, Mohamed AA, Mohammad Jawaid, Braganca R, Elias R. Development of pulp moulded packaging samples from empty fruit bunch fibre. *Pertanika Journal of Science and Technology* [Internet]. 2021 Oct 21 [cited 2024 Dec 25];29(4):2901–2912. Available from: http://www.pertanika.upm.edu.my/pjst/browse/regular-issue?article=JST-2648-2021
25. Rizal S, Olaiya FG, Saharudin NI, Abdullah CK, Olaiya N, Mohamad Haafiz MK, et al. Isolation of textile waste cellulose nanofibrillated fibre reinforced in polylactic acid-chitin biodegradable composite for green packaging application. *Polymers.* 2021 Jan 20;13(3):325.
26. Broda J, Przybyło S, Gawłowski A, Grzybowska-Pietras J, Sarna E, Rom M, et al. Utilisation of textile wastes for the production of geotextiles designed for erosion protection. *The Journal of The Textile Institute.* 2019 Mar 4;110(3):435–44.

27. Sengupta S, Debnath S. Study on needle punched jute nonwoven as an artificial medium for germination of seed: Effect of bulk density. *Journal of Natural Fibers.* 2019 May 19;16(4):494–502.
28. Leon AL, Potop GL, Hristian L, Manea LR. *Efficient technical solution for recycling textile materials by manufacturing nonwoven geotextiles. IOP Conference Series: Materials Science and Engineering.* 2016 Aug;145:022022.
29. Chmura D, Salachna A, Broda J, Kobiela-Mendrek K, Gawłowski A, Rom M. Multifunctional geotextiles produced from reclaimed fibres and their role in ecological engineering. *Materials.* 2022 Nov 10;15(22):7957.
30. Gopalsamy S, Govindharajan T. Mesoporous silica/epoxy coated cotton fabric for durable oil-water separation. *Journal of Industrial Textiles.* 2022 Jun;51(4_suppl):5973S–6000S.
31. Zahur-Uz-Zaman MD, Anjum N, Dey S, Zakir HM. An investigation of the effectiveness of recycled coconut fiber and rice husk as filter media for wastewater treatment. *JOGEE.* 2023 Apr 7;17(3):30–9.

Index

Pages in italics refer to figures and pages in bold refer to tables.

V

W

Z

For Product Safety Concerns and Information please contact our EU representative GPSR@taylorandfrancis.com
Taylor & Francis Verlag GmbH, Kaufingerstraße 24, 80331 München, Germany

www.ingramcontent.com/pod-product-compliance
Ingram Content Group UK Ltd.
Pitfield, Milton Keynes, MK11 3LW, UK
UKHW020206200926
475702UK00004B/636
* 9 7 8 1 0 3 2 7 9 6 8 7 1 *